Managing Organizational Diversity

Carolina Machado · J. Paulo Davim
Editors

Managing Organizational Diversity

Trends and Challenges in Management and Engineering

Springer

Editors
Carolina Machado
School of Economics and Management
University of Minho
Braga
Portugal

J. Paulo Davim
Department of Mechanical Engineering
University of Aveiro
Aveiro
Portugal

ISBN 978-3-319-54923-1 ISBN 978-3-319-54925-5 (eBook)
DOI 10.1007/978-3-319-54925-5

Library of Congress Control Number: 2017933070

Printed on acid-free paper

This Springer imprint is published by Springer Nature
The registered company is Springer International Publishing AG
The registered company address is: Gewerbestrasse 11, 6330 Cham, Switzerland

Preface

Managing Organizational Diversity—Trends and Challenges in Management and Engineering covers the issues of organizational diversity in nowadays organizations. Markets, day after day, are changing more and quicker than ever. All over the world academics as well as practitioners are seeking to understand how organizations manage and/or can manage the diversity of knowledge, skills, people, workforce, cultures, and approaches that they are facing day after day, in order to obtain more effective competitive advantages. They are anxious to know what trends and challenges they need to deal with in order to become competitive and act in a pro-active way. Nowadays, greater than ever before, the need to manage this diversity is one of the main aims of organizational management. Successful organizations are those which understand the importance of all their assets, namely, financial, physical, material, human, and intellectual. The management of all of them is of critical relevance to the organization.

Conscious of the importance of these issues, and in order to answer the concerns expressed by many academics, as well as executives and managers, this book looks to help these professionals to understand and implement in their organizations effective strategies, policies, and practices of how to manage organizational diversity. It looks to show what are the different trends and challenges that organizations are facing in the way how they manage and/or need to manage their organizational diversity.

Looking to communicate the recent developments and thinking in what concerns the latest research activity relating to organizational diversity management world-wide, the present book provides discussion and the exchange of information on principles, strategies, models, techniques, methodologies, and applications of the organizational diversity management, trends and challenges, in the field of industry, commerce and services.

Following these concerns, this book provides, in seven chapters, a channel of communication to disseminate, among academics/researchers, managers and engineers, the way how organizations are developing in order to maximize and increase the add value that they can obtain from their organizational diversity and the way how this diversity is managed. More precisely, Chap. 1 discusses "*Inclusion: Diversity Management 2.0*";

Chap. 2 contains information about "*Hard Times, Less Compassion? Distinct Perspectives Towards Distinct Minorities in the Portuguese Organizational Context*"; Chap. 3 covers "*Cultural Complexity in Large Organisations*", Chap. 4 describes "*Employee and Human Resource Managers Perceptions About Family-Friendly Work Practices: A Case Study Focused on Perceived Organizational Support*", Chap. 5 focuses on "*Workforce Diversity in Small- and Medium-Sized Enterprises: Is Social Identification Stronger Than the Business Case Argument?*", Chap. 6 gives information about "*Gender and Entrepreneurship in Angola: Narratives of the 'Muambeiras' of Lubango*", finally, in Chap. 7 "*Expatriate Women: A Dream Waiting to Come True*" is presented.

These theoretical and practical contributions will lead to an upper level of knowledge of these functional managerial and business subjects, at the same time that it will contribute to the acquisition of new conceptual skills able to answer to the challenges and changes set by the competitive business environment in which organizations are involved.

The Editors acknowledge their gratitude to Springer for this opportunity and for their professional support. Finally, we would like to thank to all chapter authors for their interest and availability to work on this project.

Braga, Portugal Carolina Machado
Aveiro, Portugal J. Paulo Davim

Contents

Editors and Contributors

About the Editors

Carolina Machado received her Ph.D. degree in Management Sciences (Organizational and Politics Management area/Human Resources Management) from the University of Minho in 1999, and Master degree in Management (Strategic Human Resource Management) from Technical University of Lisbon in 1994. Teaching in the Human Resources Management subjects since 1989 at University of Minho, she has been Associate Professor since 2004, with experience and research interest areas in the field of human resource management, international human resource management, human resource management in SMEs, training and development, management change, and knowledge management. She is Head of Human Resources Management Work Group at University of Minho, as well as Chief Editor of the International Journal of Applied Management Sciences and Engineering (IJAMSE).

J. Paulo Davim received his Ph.D. in Mechanical Engineering in 1997, M.Sc. degree in Mechanical Engineering (materials and manufacturing processes) in 1991, Dip Eng Graduate (5 years) in Mechanical Engineering in 1986, from the University of Porto (FEUP), the Aggregate title from the University of Coimbra in 2005 and a D.Sc. from London Metropolitan University in 2013. He is Eur Ing by FEANI-Brussels and Senior Chartered Engineer by the Portuguese Institution of Engineers with a MBA and Specialist title in Engineering and Industrial Management. Currently, he is Professor at the Department of Mechanical Engineering of the University of Aveiro. He has more than 30 years of teaching and research experience in manufacturing, materials and mechanical engineering with a special emphasis in machining & tribology. Recently, he has also interest in management/industrial engineering and higher education for sustainability/engineering education. He has received several scientific awards. He has worked as evaluator of projects for international research agencies as well as examiner of Ph.D. thesis for many universities. He is the Editor in Chief of several international journals, guest editor of journals, books editor, book series editor and scientific advisory for many international journals and conferences. Presently, he is an editorial board member of 30 international journals and acts as reviewer for more than 80 prestigious Web of Science journals. In addition, he has also published as editor (and co-editor) more than 80 books and as author (and co-author) more than 10 books, 60 book chapters and 400 articles in journals and conferences (more than 200 articles in journals indexed in Web of Science/h-index 36+ and SCOPUS/h-index 45+).

Contributors

Sandra Amorim Department of Management, School of Economics and Management, University of Minho, Braga, Portugal

Iris Barbosa Department of Management, School of Economics and Management, University of Minho, Braga, Portugal

Claudia Carvalho Department of Management, School of Economics and Management, University of Minho, Braga, Portugal

Nick G. Chandler Budapest Business School, Budapest, Hungary

Emilia Fernandes Department of Management, School of Economics and Management, University of Minho, Braga, Portugal

Carolina Feliciana Machado Department of Management, School of Economics and Management, University of Minho, Braga, Portugal

Evanice Nadya Department of Management, School of Economics and Management, University of Minho, Braga, Portugal

Deirdre O'Donovan Department of Organisational and Professional Development, School of Business, Cork Institute of Technology, Bishopstown, Cork, Ireland

Gina Gaio Santos Department of Management, School of Economics and Management, University of Minho, Braga, Portugal

David Starr-Glass University of New York in Prague, Prague, Czech Republic; Empire State College, State University of New York, Saratoga Springs, NY, USA; Jerusalem, Israel

Inclusion: Diversity Management 2.0

Deirdre O'Donovan

Abstract This chapter discusses the concept of inclusion, and proposes that organisations should move beyond traditional diversity management initiatives towards inclusion. It is not suggested that organisations skip over diversity management. Instead, it is suggested that they begin with diversity management and move towards inclusion, which, as the title of this chapter suggests, can arguably be considered Diversity Management 2.0. This chapter begins by defining and explaining the concept of diversity and diversity management, before presenting an argument for inclusion and outlining how organisations may begin their inclusionary efforts.

Diversity: What Is It?

Diversity is not a new phenomenon, rather has always been present in societies. Any two people are diverse from each other, no matter how similar they may appear. Given that diversity has always been present in societies, it is arguable that diversity has also always been present in organisations [1]. An apparent increase in diversity in workforces today may be attributable to a number of factors, including, for example, globalisation, anti-discrimination legislation and changes in demographics [2–8]. Regardless of the reason for its presence, and organisational interest in the concept, in order to better understand what diversity means for organisations, the concept must first be understood.

Although ever-present, defining diversity is difficult. First, difficulties lie in the fact that diversity concerns both visible and invisible characteristics [8, 9], and can be considered subjective, meaning it is created by individuals who characterise others as similar or dissimilar to themselves [10]. Diversity can also be considered context dependant, as individuals cannot be defined as "different" in isolation,

D. O'Donovan (✉)
Department of Organisational and Professional Development, School of Business,
Cork Institute of Technology, Rossa Avenue, Bishopstown, Cork, Ireland
e-mail: deirdre.odonovan@cit.ie

C. Machado and J.P. Davim (eds.), *Managing Organizational Diversity*,
DOI 10.1007/978-3-319-54925-5_1

rather only when compared to others in their environment [9]. Diversity is also a relative term [9]. The relativity of diversity considers diversity indicators as ambiguous as, for example, two individuals may identify as female, but one may be more "feminine" than the other.

In the organisational context, Jackson and Joshi [6] develop the concept of diversity further by exploring workplace or work team diversity. In doing so, the concepts of Relations-Oriented Diversity, Task-Oriented Diversity, Readily Detected Diversity and Underlying Diversity were introduced. Relations-oriented diversity concerns attributes that are instrumental in shaping interpersonal relationships, but have no ostensible implications for the performance of tasks. These attributes include, gender, age and religion. Task-oriented diversity concerns attributes, such as for example, education level, experience and cognitive abilities, which do seem to have an implication for work task performance. Readily detected diversity concerns differences among team members on a number of characteristics which are relatively easily discernible, such as age or nationality. Underlying diversity, in contrast, concerns attributes which become evident through interaction, such as personality or attitudes [6].

It is important to note that while diversity studies traditionally focussed on diversity in terms of gender, ethnicity and culture, diversity encompasses many other differences, perhaps almost an infinite number [11]. Although difficult to define, a number of definitions of diversity exist. Joplin and Daus [12] offer a clear, simple, concise definition of the complex concept, proposing it to simply refer to all the ways in which people differ from each other. Although brief, this definition underscores that individuals can differ from each other in a multitude of ways.

Under a more workplace-oriented focus, Griffin and Moorhead [4: 31] define diversity as concerning:

> The similarities and differences in such characteristics as age, gender, ethnic heritage, physical abilities and disabilities, race and sexual orientation among the employees of organizations.

However it is defined, diversity is a broad, complex concept that concerns every individual. Additionally, diversity in the workplace is considered to carry implications for management, and so must be managed. Consequently, diversity management is more fully discussed later in this chapter.

Benefits Associated with Diversity

Diversity in the workplace can offer organisations a number of potential benefits. According to a number of authors, leveraging diversity in the workforce is increasingly being seen as a strategic resource for competitive advantage [13–15]. It has also been argued that diversity is critical to the success of an organisation's bottom line [16]. It is interesting to note that many of the advantages associated with diversity mirror those associated with inclusion, as will be evident later in the chapter.

One potential advantage associated with diversity concerns Cost Savings. Cost savings, in this instance, focus on the negative impact the mismanagement of diversity has on an organisation's bottom line. This negative impact specifically refers to higher staff turnover costs, higher absenteeism rates and lawsuits on sexual, age and race discrimination. Regarding higher turnover costs, turnover among diverse employees is a costly and significant problem for many organisations, as are the subsequent added recruiting, staffing and training costs per person. Additionally, a persistent flow of employees through an organisation results in employees continually climbing the learning curve, rather than performing to their full potential [13, 17]. It is arguable, therefore, that managing diversity enables employees to perform to their potential [18].

The second element of the cost savings argument concerns lowering absenteeism rates. Absenteeism rates, as has been well documented, can amount to significant costs for organisations. Absenteeism can occur when individuals do not feel secure about their status, as such insecurity prevents employees from fully engaging at work. Monks [7] suggests that the introduction of diversity initiatives has a positive influence on absenteeism rates, tending to result in a reduction of both labour turnover and absenteeism levels.

The final aspect of the cost savings argument focuses on lawsuits on sexual, age and race discrimination, or, perhaps more specifically, a strategic organisational effort to avoid their occurrence [17, 19, 20]. Diversity programmes should assist organisations in complying with laws regarding discrimination, and ensure that policies and processes are in place in organisations to deter discrimination lawsuits, as organisations that are conscious of the diversity of their workforce are more likely to anticipate problems, thus potentially reduce the risk of litigation [13, 21]. Kim [22] examines this argument from an alternative angle, discussing it in terms of company image. Organisations can focus on improving the company's public image or enhancing its image by using diversity management to reduce the chance of discrimination law suits. It must be cautioned, however, that with the exception of costs relating to turnover, actual cost savings from improving diversity management are difficult to measure [23].

Winning the competition for talent, or the "talent war", refers to the attraction, retention and promotion of employees from different demographic groups, and is another argument for managing diversity [14, 17, 24]. It has been strongly suggested that an organisation's future is dependent on the quality of talent it attracts and retains [16]. To sustain a competitive advantage, organisations must be able to optimise their human resources [17]. Indeed, Carbery and Cross [25] contend that for many organisations, one way in which effectiveness is measured is via the achievement, and maintenance, of sustainable competitive advantage. Organisations should focus on identifying, retaining and developing their key employees in order to gain a competitive advantage over competitors [26]. Organisations that are able to recruit, develop, retain and promote diverse employees are more likely to have an edge over their competition, as talented employees will be attracted to organisations that value their capabilities. These employees will also be more willing to invest in productive activity if they believe they are being treated fairly, and that career opportunities are available to them [17]. Espinoza [13] similarly advocates a belief that a diverse workforce allows diverse employees to identify with the company, making the company attractive to other diverse potential employees.

Diversity can also aid in employee retention, as commitment to diversity indicates to employees that the organisation cares for them as individuals [13, 21]. Organisations are cautioned that if they do not effectively manage diversity issues, diverse talent will leave in favour of a competitor who does [27, 28]. That being said, it is also noteworthy that Schneider's [29] attraction-selection-attrition (ASA) model indicates that the attraction and retention of diverse employees may not be a straightforward process, arguing that organisations naturally evolve towards social homogeneity as individuals prefer to be with others who bear similarities to them. While candidates are more attracted to organisations that they believe are made up of individuals similar to themselves, if, once hired, these employees do not seem to fit in with the rest of the organisation, they are more likely to experience dissatisfaction and, ultimately, leave. If such a condition repeats over a period of time, the result is a gradual homogenisation of an organisation [10, 29]. While this seems to suggest that homogeneity of the workforce may be inevitable, to avoid this natural drift towards homogeneity, and subsequent increased turnover costs, Jackson and Joshi [6] indicate a proactive approach to increasing diversity may be necessary. Arguably, this position would appear to support the assertion that diversity must be actively managed.

Driving business growth is another argument in favour of managing diversity [17]. Driving business growth centres on organisations managing diversity to leverage a number of possible opportunities. The first opportunity is based on organisations using workforce diversity to gain an increased understanding of the marketplace in which they operate [21, 27, 30, 31]. A 2003 report on diversity undertaken by the European Commission cited improved access to new market segments and improving performance in existing markets as benefits of diversity [30, 31]. Furthermore, customers and suppliers are becoming increasingly diverse, as indeed is the marketplace as a whole [16, 17, 23, 32]. It arguably makes sense that the understanding needed to market to diverse demographics, and to respond to their needs, naturally resides in marketers with the same background [17, 33]. Additionally, it has previously been found that individuals from a minority culture are sometimes more likely to give patronage to a sales representative from their own culture [23]. Similarly, in addition to gaining market penetration, organisations can benefit from the goodwill of diverse customers who prefer to buy products produced by a diverse workforce, or who prefer to do business with organisations that have a diverse sales force [17]. Espinoza [13] advocates this line of reasoning, believing that an organisation's sales force should match its customer base, adding that diversity provides a good image to an organisation's customer base, and enhances company branding.

A further opportunity arising from diversity is greater employee creativity and innovation [7, 17, 27, 31, 34–36]. Attitudes, cognitive functioning and beliefs tend to vary with demographic variables such as gender, age and race. As diversity shapes how we view things, one consequence of diversity in an organisation's workforce, therefore, is the presence of different perspectives or views on the performance of tasks [16, 17, 34]. If the varying approaches, views or opinions are considered, the likely result is the enablement of management to make better and more informed decisions [13]. Additionally, managing diversity can make employees feel valued and supported, which tends to result in employees becoming more innovative [37]. It is also suggested that diversity can increase the quality of team problem-solving, as diversity among team members enables employees to see problems from an array of perspectives, based on their wide range of experiences, potentially producing better decisions [17, 21, 23, 24, 35, 38]. Van Knippenberg [38] cautions that in reality, however, groups often struggle in harnessing the potential advantages of diversity, which is perhaps unsurprising given our knowledge of the issue of group-think and the notion of group culture as a subculture of organisational culture. Additionally, organisations must now also attempt to not solely source the best individual for a position, but also consider the best combination of individuals in terms of their characteristics [10].

In addition to enhancing group and individual performance, less emphasis on employee conformity to past norms, via the open acceptance of diversity, should also improve creativity [23]. The notion that innovation is a positive consequence of the presence of a diverse organisation or team is grounded in two propositions [10]. First, it is assumed that diverse individuals have diverse, and consequently more novel, ideas. Second, if individuals approach the same task from diverse points of view, task-related conflicts are more likely to occur. Dealing with these conflicts should result in a more thorough consideration of all aspects and approaches, culminating in more innovative solutions [10, 39]. A related argument for managing diversity relates to the earlier mentioned concept of group-think, or rather avoiding it. Conflicts due to diverse perspectives result in questioning, and moving beyond, prior practices, thus require questioning of current ideas or practices and the overcoming of group-think [10].

Diversity in organisations offers the potential to improve effectiveness at higher levels in the organisation. Heterogeneity of top management can prevent a myopic perspective at senior levels, and so leveraging diversity in higher levels of the organisation can provide the organisation with an opportunity to improve leadership effectiveness. Furthermore, the increased awareness developed by organisations that manage or adapt to diversity can help them become more effective in cross-cultural business situations [17, 21, 22, 27]. Additionally, good diversity skills are compatible with good people management skills, and so focussing on management's ability to supervise a diverse workforce can result in improvement of their overall people management skills [21].

Managing diversity can also improve the organisations bottom line [2, 13, 14, 40–42]. Employees who believe their employer supports them have a tendency to be more productive. This increase in productivity positively impacts the organisation's bottom line [2, 13]. A commitment to diversity enables every employee to contribute their individual ideas, talents and skills to the organisation, which again ultimately drives the organisations bottom line [40]. Moore [9], however, argues that the link between diversity and performance is not automatic or straightforward.

As well as potentially improving organisational performance, diversity can also impact an organisation's flexibility. Through managing diversity, organisations should become less standardised, and more fluid. This fluidity should, in turn, create greater flexibility, enabling organisations to react to environmental changes with greater speed, and at less cost [23]. Alternatively, organisations used to offering flexible arrangements, such as family friendly or work–life balance opportunities, may be better placed to overcome skills shortages or provide alternatives to redundancies in difficult times through career breaks or job-sharing initiatives, thus retaining their employees [21]. The culture of presenteeism in many organisations, however, can result in limited uptake of flexible working arrangements by employees due to fears that using such arrangements may result in reduced career opportunities, thereby nullifying the potential benefits [7].

While a number of arguments have been presented for diversity management, there are also many potential challenges associated with diversity in organisations, as discussed in the next section of this chapter.

Challenges Associated with Diversity in the Workforce

Managers may now find themselves with a new and pressing, or sensitive, set of challenges that were not as dominating, or perhaps were irrelevant, in an (apparently) homogenous workforce [12, 36]. Moreover, although managers are being increasingly called upon to deliver diversity strategies, there is little evidence that managers are receiving the training or support necessary to do so [7]. Consequently, managers may be more likely to view diversity a marginal activity, and be reactive, concentrating on minimal compliance, rather than proactive, concentrating on possible positive outcomes.

The genuine support and commitment of top management to diversity is crucial [7, 23, 28, 43]. A lack of commitment on the part of top management may pose a challenge, because if they do not talk about diversity, and embrace its values, diversity initiatives will not work [13]. Resources, such as human, financial and technical, must be committed, and provided to the organisation's diversity initiatives [23, 44]. Indeed, commitment from the organisations top leadership is seen as a component of a best practice approach to diversity management by the U.S. government's Accountability Office [45]. While crucial, however, top management support alone is not sufficient. The use of champions for diversity at lower levels in the organization, and employee involvement in driving diversity, is also of great importance [23, 28, 43].

A significant challenge to diversity arises if various groups believe diversity is only important to their group. If these groups have their own separate agendas, rather than working together to improve diversity throughout the organisation, top management may not believe in the merits of diversity [13]. A further issue concerning groups relates to grouping individuals based on generalisations. Organisations have, for example, a tendency to treat female employees and ethnic minorities as homogeneous groups. Doing so results in neglecting individuals in those groups who do not fit the profile of the stereotypical member of that organisationally formed group [46]. Doing so also neglects consideration of all the ways in which the individuals in that group differ from each other.

Fully accepting diversity means accepting change about how business is done. This forms another challenge, as many individuals are uncomfortable with change, therefore resist it [13, 18, 44, 47]. Diversity management should bring about a change in recruiting and retention policies, as well as a change in how people view and accept differences [13]. A challenge is also posed by cosmetic changes, which disguise what really happens in the organisation.

The fear of reverse discrimination has also been raised as a challenge to managing diversity. Some employees may believe that managing diversity is a smokescreen for reverse discrimination, and so these employees may resist diversity management initiatives [18, 20]. Carnevale and Stone [2] and [48] also highlight reverse discrimination as a challenge, proposing that men in particular, specifically white men, are being forgotten about by organisations. Such fears are reflected in the Resistance Paradigm for managing diversity. This paradigm develops through concerns by a majority that they may be displaced by minorities. Under this paradigm, all visible differences, and increasing pressure for diversity, are considered threats [49]. As such, it is important that white males are included in the organisations future vision for diversity, and their role in achieving such is clearly outlined [2]. Indeed, also referring to men, Muzio and Tomlinson [50] and Smithson and Stoke [51] note that work–life balance policies are often seen as policies for women's problems, even though such policies are gender neutral. Similarly, Kent and O'Donovan [52] highlighted that literature concerning work–life balance indicates a bias towards women.

It was earlier noted that diversity in opinions and ideas is considered a positive reason for managing diversity [16, 17, 23]. In contrast, it has also been identified as a challenge, especially for managers [12]. One reason for this negative view centres on organisations attempting to set agreement on important matters. Previously, such agreements may have been quickly achieved, whereas now, managers in diverse organisations now have to sift through, and decipher, a number of different perspectives on the same problem or issue. This can be a rather time-consuming process, and the task of management becomes more complex [36].

Increasing interaction among diverse members in the organisation also increases the potential for friction in the organisation [2, 53]. Friction and resulting tension can reverberate throughout the organization, causing a reduction in productivity, an increase in costs and reduced quality products or services. Tensions may also arise as a result of culture clashes. Culture clashes can be a drain on the performance of individuals involved, and perhaps undermine or damage organisational culture. Consequently, work relationships and output may suffer [2]. Indeed, there is, on occasion, an overall pessimistic view of diversity which suggests that diversity creates social divisions, resulting in negative outcomes for the organisation [54].

Tokenism, whether real or perceived, can present a further challenge for organisations [12]. Tokenism occurs when an individual is hired over more qualified candidates, either in an effort to address the concerns of stakeholders, or to fulfil quota numbers. While quota systems are rarely in an organization's best interests, in an organization that has little tolerance for diversity, quotas may be the only way to ensure that diverse candidates will be included in recruitment and selection processes [12]. The use of quota systems, through which organisations focus their recruitment and selection activities at particular diverse groups, is advised against by a number of authors, for a number of reasons. Joplin and Daus [12] and Von Bergen et al. [20] believe quotas automatically result in a perception of tokenism. Flynn [48] warns that quotas can lead to discrimination towards white men. Perceived tokenism often occurs when the diversity of an organization is increased, as growing diversity often carries the perception that less qualified candidates are being hired. In addition to the perceptions of existing employees, new employees who believe they were hired for anything other than their merit may become defensive, feel vulnerable and eventually begin to question their capabilities [12]. These issues may result in increased turnover rates among new hires. The use of quotas may result in organisations attempting to manage diversity through the Discrimination and Fairness Paradigm, although perhaps not consciously. This paradigm is based primarily on equal opportunities, fair treatment, recruitment and compliance with legislation, but the potential difficulty is that it tends to focus too much on achieving what is perceived as the "right number" of diverse employees [22], rather than the right people for the job. Thomas and Ely [55] do observe, however, that while organisations operating under this paradigm measure progress in diversity by how well they achieve their recruitment and retention goals, it does actually move beyond being solely concerned with numbers.

Finally, while cost savings was previously discussed as an advantage associated with diversity, it is also possible that companies investing in diversity face additional costs. Organisations may face costs associated with legal compliance, cash costs of diversity, opportunity costs and business risks [30]. Potential costs associated with legal compliance may include employee training, record-keeping processes and the cost of communication of new policies. The extent of these costs for different companies will be influenced by the nature of existing internal processes and legislative requirements. Cash costs associated with diversity may be short term, "one-off" costs, for example, improving access to buildings for employees with mobility impairments, but are often long-term and recurring. Potential cash costs include, for example, those associated with the necessity for specialist staff and the provision of training, the provision of support and facilities, communication costs, the development of employment policies and monitoring and reporting processes. Opportunity costs associated with diversity may include managers' time and productivity shortfalls. The business risks of diversity centre on the tendency for many programmes which have been designed to change organisational culture taking longer than intended, or failing. This phenomenon is referred to as execution risk [30].

Regardless of the potential challenges associated with diversity, given the potential advantages and the ever-presence of diversity, it must be managed. In addition, this chapter proposes that diversity management is a gateway to inclusion initiatives, and so managing diversity is discussed in the following section of this chapter.

Diversity Management: What It Is, and How to Do It

Different people behave in different ways [56]. Employees who may appear similar, perhaps with regard to age or gender, are still very different individuals, and so may respond differently to the same stimulus, and various management styles [57]. As such, rather than treating every employee in the same way, managers must instead recognise and respond to employee differences in such a way that retention and productivity are maintained, while discrimination is avoided and fairness is maintained. This, Robbins [58] suggests, is one of the most important challenges facing organisations today.

As discussed earlier, diversity may present organisations with a number of benefits. The mere presence of a diverse workforce is insufficient to realise these advantages [59]. Instead, to leverage these benefits, organisations must manage diversity [23]. It should initially be noted that managing diversity and valuing diversity are not the same. Valuing diversity is a more passive phenomenon, referring to being aware of the relative worth and importance of diversity, whereas managing diversity is an active phenomenon, which involves coordinating and directing the differences of employees to ensure strategic organisational goals are met [60]. With that in mind, however, it is also noteworthy that Liff [46], who has also distinguished between valuing and managing diversity, does suggest that valuing diversity is a possible version of diversity management, albeit one that lacks a strategy for overcoming potential problems associated with diversity.

Managing diversity does not mean controlling or containing it, nor does it involve the assimilation of employees to fit the organisations existing culture [21, 61]. Instead, diversity management refers to a concept of enabling each member of a workforce to perform to their potential [61]. Such enablement requires organisations to adopt a new way of thinking about differences among people, and a new approach to the way in which employees are treated [18, 21]. Additionally, as referred to in the previous section of this chapter, in order to be truly successful, diversity management must receive support from the top levels of the organization [2, 13, 14, 28, 44, 49, 62]. It is essential that managers take a proactive approach to their involvement with employees [12]. One method by which top management commitment may be secured is to involve managers in the planning process, giving them ownership of diversity goals [2]. It is also arguable that developing a diverse organization necessitates the presence of diverse management [63].

Organisations are cautioned that it is ill-advised to make diversity management the responsibility of a single individual, as diversity management initiatives may collapse, if and when, that individual leaves the company [28]. To offset this concern, diversity management should be made a company-wide issue. It is also important to note that as everyone, regardless of superficial similarities, has a different background, some will emphatically embrace diversity, others will not be sure about it, and others will simply view it as a nuisance [13].

There is no single best way to create a model of diversity management, as each organisation is different [21, 62]. IBEC [21] propose a four-phased cyclical diversity management model, as seen in Fig. 1, which may be adapted to meet an organisations particular need.

The first stage, analysis, involves constructing a profile of the level of diversity that exists among the organisations workforce and customer base. Following this, existing or current employment processes must be examined to identify areas relating to diversity that require addressing [21]. Kochan et al. [64] are also advocates of such analysis, arguing that regardless of the widespread use and availability of Human Resources information systems, they have found that HR data concerning groups and individuals could not be readily linked to performance. Consequently, HR practitioners, and organisations, are limited in terms of learning how to effectively manage the diversity in their company, weakening their stance on the strategic importance of diversity.

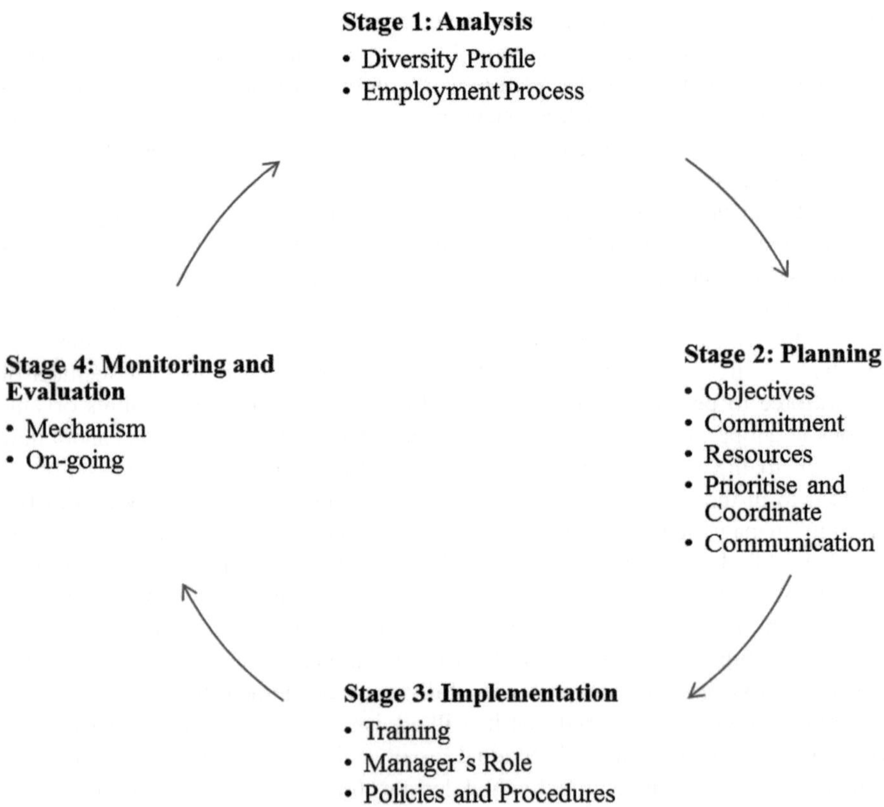

Fig. 1 IBEC's diversity management model. *Source* O'Donovan [1], Adapted from IBEC [21: 8]

The second stage requires organisations to take a comprehensive approach to establishing the key objectives behind the diversity strategy. Following this planning stage, the diversity management programme should be ready to implement, bringing the organization to stage three in the cycle. Successful implementation requires diversity strategies to not solely be seen as a human resources issue, rather, as vital to the achievement of the organisations goals. Additionally, while it may sometimes be appropriate to devise new policies, amendment of existing policies and procedures is also important. Once the programme has been implemented, it is crucial that it is regularly assessed, and, if found to not be achieving the desired objectives, amended, requiring movement through the cycle again [21].

Many executives may be unsure of why they should want to learn how to manage diversity [61]. Moore [9] makes reference to a particular perspective of diversity in organisations termed Diversity Blindness. This perspective does not consider diversity an issue that must be addressed, resulting in diversity and diversity training being ignored. Consequently, organisations proposing to manage diversity should begin by clarifying their underlying motivation, as, given today's competitive challenges, it is likely that only business reasons, specifically reasons that highlight potential improvements in the organisations bottom line, for example those outlined in this chapter, will result in the long-term motivation critical to managing diversity [61]. Indeed, a report undertaken on behalf of the European Commission proposes that companies adopt policies for three types of reason; ethical, regulatory and economic, or a mix of the three [30].

Doke and Beagrie [41] propose that the starting point of any diversity management programme is to communicate to an organisation's workforce what diversity is, what the organization is aspiring to achieve, and the goals the organization has set in place to assist it in achieving its aim of effective diversity management. Indeed, Miller and Tucker [43] highlight that creating awareness of benefits of diversity among management and employees is important, as doing so increases commitment to furthering the diversity goals of the organization. Kreitz [44] agrees, and suggests that human resources directors and senior managers should express the motives behind their interest in diversity, and identify the ways in which diversity will benefit the organisation. A number of steps may be taken by organisations to communicate the reasons for managing diversity to its employees, and to initiate the process of diversity management.

First, it is essential that the vision is clarified [41, 61]. The ideal vision to be communicated to the organisations employees is an image of fully tapping the human resource potential of each individual in the workforce. Additionally, managers must attempt to expand their focus [61]. There is a tendency for equal employment opportunities to focus on women and minorities, offering little to white men in particular, who are just as diverse in numerous ways. Indeed, [48] cautions that white men may see a negative side to diversity programmes, as they are being grouped into one bundle by some.

As the goal of diversity management is to create a heterogeneous culture, or to make use of their heterogeneity, organisations should undertake a corporate culture audit. As a corporate culture comprises a collection of unspoken and unexamined values, assumptions and mythologies, a culture audit is impossible to conduct without external assistance [61]. Cox and Blake [23] are advocates of this step, being of the opinion that a comprehensive analysis of organisational culture and organisational systems such as recruitment, performance appraisal, assessment of potential and promotion and compensation should be conducted. Organisations can use this audit to uncover sources of unfavourable potential bias towards, or against, certain groups, and to identify ways in which the organisations culture may inadvertently put some members of the organisation at a disadvantage.

Carrying on from their cultural audit, organisations need to modify their underlying cultural assumptions. A particular problem with corporate culture is that when changes to it are attempted, they are met by intense opposition [61]. Regardless, organisations must still attempt to modify their cultural assumptions if they are to succeed in transforming their organisation from homogenous to heterogeneous. The first purpose of modifying underlying cultural assumptions is the enablement of organisational systems reform. Organisations should identify not whether the system is at maximum efficiency, rather whether the system works for all employees. The second purpose of assumption modification is to modify models of managerial and employee behaviour, as managers seek subordinates who will do as they do, or subordinates who aspire to be like them [61].

Organisations also need to assist their employees in becoming pioneers of diversity. Learning to manage diversity constitutes a change process, thus, the managers of the organization must become change agents. Top management articulates the organisations new diversity policy and their commitment to it, yet it falls to middle management to implement the policy, and deal with any new resulting problems. To help them do so, these managers should be appropriately trained, and reminded of their importance in diversity efforts.

The penultimate step requires organisations that aim to manage diversity to apply a special consideration test to diversity programmes. A number of questions are to be addressed, specifically:

- Does the programme/principle/policy give special consideration to one group?
- Will the programme/principle/policy contribute to everyone's success, or just the success of one group?
- Is the programme/principle/policy designed for them as opposed to us? [61: 115].

If the answer to these questions is yes, the organisation is not yet on the way to managing diversity.

Finally, organisations who are trying to learn how to manage diversity can continue to use affirmative action. Similar to the usage of quotas, organisations do need to move beyond affirmative action, as affirmative action does not deal with the causes of prejudice and inequality, nor does it help to develop the potential of every individual in the organisation. An alternative final step has been suggested by Cox and Blake [23], termed "follow-up". Follow-up consists of monitoring change, evaluating results and institutionalising the changes as part of the organisations continuing processes. Follow-up should include more training and repetition of the audit step. Organisations can also use focus groups to facilitate continuing discussions on diversity issues. Alternatively, organisations can, as part of their greater diversity management initiatives, engage in diversity training.

Regardless of the approach to diversity management employed, there is a tendency to become simplistic or rigid when considering differences [65]. Ferdman and Brody [66] argue that differences should not be the basis for invidious distinctions between individuals, rather should be a source of pride, and used to the benefit of all. It is necessary, therefore, that diversity management go further than complying with existing legal rules or attempting to react to shift in workforce demographics [67]. Organisations need to move beyond diversity management towards creating an organisational environment that is inclusive for all employees [68]. Individuals need to feel, and be, included in their professional workplace environments [69]. Indeed, Anderson [34] proposes that any diversity initiative will be more successful if managers engage and use processes that foster equity, consensus and empowerment among, and of, employees, while Sabharwal [68] notes that many authors have articulated that inclusion is the crux of organisational diversity efforts. Ultimately, organisations should move towards inclusion.

Inclusion

Many individuals consider their individuality a significant part of themselves, a part which they would not like to be overlooked [70]. In addition, to realise the potential benefits of diversity, it is insufficient to simply hire and retain diverse employees, rather, these diverse employees must be more fully integrated into the social fabric of the organization [71]. Increasing emphasis is being placed on the need to leverage multiculturalism and foster inclusion as a basis for the success of an organization [72]. Pless and Maak [67] propose that organisations who take an assimilation approach to diversity, which largely ignores differences, rather than an approach of integration and inclusion, will struggle to achieve the potential benefits afforded by a diverse workforce, as discussed earlier. Similarly, Ferdman [59] states that it is not the presence of diversity, rather how it is addressed, that leads to positive outcomes. Before proceeding, it should be noted that diversity and inclusion, although related, are not interchangeable terms for the same concept, rather are separate [73].

When discussing how diversity should be "done" in organisations, Davidson and Ferdman [74] propose that the answer rests in an inclusive version of diversity. An inclusionary approach to diversity management is one in which:

> Differences are recognized, valued and engaged. Different voices are understood as being legitimate and as opening up new vistas; they are heard and integrated in decision making and problem solving processes; they have an active role in shaping and fostering creativity and innovation; and eventually in adding value to the company's performance [67: 130].

Ferdman [59: 4] concisely notes that:

> Inclusion involves how well organizations and their members fully connect with, engage, and utilize people across all types of differences.

Based on these definitions, under an inclusionary approach to diversity management, the differences of individuals in the workplace are not just identified, rather are integrated into the fabric of the organisations culture. Consequently Pless and Maak [67] propose the argument that to fully realise the potential of diversity, a culture of inclusion must be developed and established by the organisation. A culture of inclusion fosters enhanced integration amongst employees, and activates latent diversity potentials. Such a culture, built on clarified normative foundations, honours both the differences of employees, and their similarities [67]. Full inclusion, and truly valuing differences, requires the implementation of organisational processes that involve all members of the community, or organization [65, 74].

The Nature, Dimensions and Levels of Inclusion

Inclusion is a contextual, individual, momentary and transient concept, which occurs at both the organisational and individual level [1, 59]. Organisations should also be aware that employees can feel simultaneously included and un-included. Individuals may perceive inclusion in one area, perhaps among the colleagues in their team or department, but not in the wider organization. It does not follow that an employee who feels included in one context will perceive inclusion in all other organisational contexts, and it is not automatic that an individual perceiving inclusion now will do so in the future. This is significant, as it means that organisations that are aiming to be inclusive, indeed, may generally be considered to be so, may have individuals in the workforce who do not perceive inclusion. Organisations are therefore cautioned that fostering inclusion at the individual level is an ongoing process. As such, organisations should operate under the hypothesis that inclusion is not a static concept, with employees perceiving or not perceiving inclusion, rather, it is contextually dependant. Consequently, organisations are advised to make an effort to monitor inclusion on a continual basis, to ensure that employee's perceptions of inclusion remain, even when operating in another context, for example, on a different team or in another department [1].

Furthermore, developing inclusion just at an individual level is likely insufficient, cognisant that different departments may have different cultures and manners of operating, meaning, as mentioned above, an individual may feel included in one area, but not in another. If there is an attempt to be an overall inclusive organisation, this may reduce the likelihood that individuals feel un-included when outside of their department [1]. In addition, some individuals may believe they need to take an active role in developing their feelings of inclusion, yet others may believe it occurs naturally [1, 59]. Organisations are again reminded, therefore, to examine inclusion efforts at both an overall organisational level, an individual level, and also at a team level. Doing so will assist in creating an organizational environment to support individuals, who believe inclusion should naturally occur, while also assisting in maintaining an individual's perception of inclusion even when operating in different departments or with different functions [1].

Inclusion is multi-dimensional. Indeed, Nishii et al. [71] propose three dimensions of organisational inclusion, specifically, Foundation of Fairness in Employment Practices, Organisational Culture of Inclusion, and Inclusion through Participation. The Foundation of Fairness in Employment Practices dimension relates to the extent to which an organisations HR policies ensure a level playing field for all employees. To create a wholly inclusive environment, organisations must design and implement practices without bias to ensure both diversity throughout the organization, and goodwill on the part of employees. While the presence of employee-friendly policies is important, the presence of such policies, however, does not necessarily translate into the development of an inclusive organization [68]. It was noted earlier, for example, that family-friendly work–life balance policies often appear geared towards women. Indeed, in a study by Kent and O'Donovan [52], it was noted that in participating organisations, more flexible work options were available to women.

Dimension two, Organisational Culture of Inclusion, refers to the extent to which the organisations basic assumptions, values and norms are inclusive of all employees. If they are truly inclusive, employees do not feel a pressure to conform to an ideal employee profile. Consequently, employees do not feel a pressure to hide or face a conflict with their identity [71]. Truly inclusive organisations help those employees who feel they do not belong to the mainstream feel part of the organization as a whole, while also helping those who do feel they belong to continue to feel so [74]. Under the dimension Organisational Culture of Inclusion, organisations devote resources to equipping their employees with the skills necessary for cooperation with other diverse individuals, recognise different perspectives and create a culture of openness through which employees can learn from each other's views [71]. This would certainly be useful for organisations aiming to become, and continue to be, a learning organisation.

The third dimension, Inclusion through Participation, concerns the extent to which an organisation successfully capitalises on, and leverages, the diversity of its workforce, to apply learning from diverse perspectives to decision-making. This dimension is grounded in the premise that diversity only benefits an organisation when employees are encouraged to manifest their diversity on idea generation and decision-making. Rather than simply telling employees they are free to make suggestions and be involved in decision-making, employees should be actively encouraged to come forward, perhaps, for example, by being invited to attend meetings. Successful operation of this dimension necessitates both formal and informal participation. Examples of formal participation include representation on key decision-making bodies or committees, while informal participation concerns participating in every-day, on-the-job decision-making. In addition to encouraging the expression of diverse opinions and experiences, they must also be incorporated into decision-making to ensure employees feel respected and included [71]. This means that when employees are invited to contribute, and that contribution is useful, it should be employed. Failing to make use of employee involvement will likely eventually result in employees no longer coming forward with suggestions, setting back the progress of inclusion efforts. Indeed, Sabharwal [68] posits that individuals feel accepted and secure in the organisation when they are part of the decision-making process.

If an organisation falls short on any of the three dimensions of inclusion identified above, then obstacles to the full utilisation of diversity remain. Essentially, Nishii et al. [71: 2] expect that:

> The relationship between diversity and performance…will be moderated by these three dimensions: in units/organisations that score high on these dimensions, there will be a stronger, positive relationship between diversity and performance than in units/organisations that score low on these dimensions.

Similar to Nishii et al.'s [71] multi-dimensional view of inclusion, Gasoreck [75: 27], while describing inclusion at the multinational firm Dun & Bradstreet, also takes a multi-faceted view of inclusion, deeming it to concern the degree to which:

- Employees are valued, and their ideas are both taken into account and used
- Employees partner successfully both in and across departments
- Current employees feel they belong in the organization, and prospective employees are attracted to the organization
- Employees feel committed to each other, the organisation, and the organisations goals and
- The organisation continues to foster flexibility and choice, and attends to diversity.

It is interesting to note that this last point (above) arguably supports the contention of this chapter that inclusionary efforts should begin with diversity management.

While there are clearly commonalities concerning what constitutes inclusion, for example feeling valued or respected, individuals perceive these themes in different ways, therefore, even if an organization is deemed to have an inclusive culture, some employees may still not feel included [1, 69]. The core proposition, as discussed earlier, is that inclusion occurs at two levels: the individual and the organisational level. Furthermore, an individual's diverse make-up may impact their perceptions, or whether they experience feelings, of inclusion. Individuals, for example, who are introverts, may experience inclusion via the establishment of one or two social connections, while more extroverted individuals may have to interact with a larger portion of the community to feel fully part of it. Consequently, individuals are cautioned to consider that treating others as they would like to be treated may not serve to make others feel included, rather may appear to be an imposition of their own values on to others. It is important, therefore, that organisations attempt to uncover the needs of their employees with regard to what will result in feeling of inclusion, and subsequently aim to address those needs [69].

One simple method for uncovering what inclusion looks and feels like for different employees is to simply ask them [76].

Addressing individual inclusion at the individual level is not, however, as mentioned earlier, a sufficient strategy for the nurturing of an inclusive organisation. Nurturing organisation-wide inclusiveness instead requires systematic and proactive efforts at the organisational level [49, 69]. Indeed, Thomas and Ely [55] (adapted by Davidson and Ferdman 2002: 83) propose that to fully utilise, and learn from, workforce diversity, organisations must satisfy a number of preconditions, specifically:

- The organisations leaders must understand that diversity includes different opinions, insights and approaches to work
- Leaders must also understand that diversity presents both opportunities for new learning, but also challenges in the form of a need for unlearning and relearning
- Each individual should be held to high performance standards, and the organisational culture must encourage employee development through training and education. Employees must also feel valued to ensure they contribute high levels of performance
- Constructive conflict, open communication and tolerance for dialogue must be encouraged
- A non-bureaucratic process must be in place to enable employees to constructively challenge current operating methods, and reshape past policies and practices, to be a more inclusive, empowering organization.

Regardless of these proposed preconditions, it is also insufficient to solely focus on inclusion at an organisational level. Instead, both the individual and organisational levels should be viewed as independently vital for inclusion, but also interactive. Additionally, although there are a number of things organisations can do to foster an inclusive work environment, inclusion is, in many ways, a momentary creation, dependant on the particular individuals and situations. Creating an inclusive culture, therefore, is a continuous process [69].

In addition to considering inclusion at the individual and organisational level, organisations may also need to consider the concept of subordinant and dominant groups. While employees are unique individuals, they also share group membership with others as part of their identity. This has an impact both on the way individuals treat, and are treated by, others [70, 77]. Groups do not hold equal status; some are typically systematically privileged while others are typically disadvantaged. Subordinant groups are those in lower power positions, while dominant groups are those typically in a higher power position. When an individual from a subordinant group occupies a position of assigned power in an organisation, they may have significant power as an individual, however, would still be a member of a subordinant group. Conversely, an individual from a dominant group operating in a position of relatively little power in the organization may have little power as an individual, but, as a member of a dominant identity group, still likely experience benefits in both subtle and overt ways [77].

The primary implication of the distinction between, and presence of, these groups, is that well-intended efforts to create an environment of inclusion may be hindered if the dynamics of the group power relationships are not considered [77]. By looking through the lens of subordinant and dominant groups, organisations can enhance their understanding of what is required to create and participate in an inclusive organization [78]. Having been made aware of the presence of such distinctive groups, employees may be enlightened as to how members of these groups can support organisational efforts to create a culture of inclusion. Regarding dominant groups, Davidson and Ferdman [77] advance a number of prerequisites for supporting inclusion, specifically:

- Dominants should assume a position of inquisitive probability, which requires them to accept that they are a member of a dominant group, which has implications for the manner in which they engage with individuals in subordinant groups.
- Individuals from dominant groups should be encouraged to develop skills enabling them to distinguish impact from intent, in that a dominant's behaviour towards a subordinant group member may be benevolent in intention, but injurious in perception.
- An effort should be made to learn about the experiences of subordinants to assist in increasing the overall sense of organisational inclusion.
- Dominants should aim to use their positions of power and privilege to change the structures and systems that exclude or discriminate against subordinants.

It is important, however, that dominants are not made to feel that they are "the problem". Indeed, subordinants have a role in the fostering of an organisational culture of inclusion, albeit rather different. It is recommended that subordinants:

- Assume a position of cautious openness, by engaging in dialogue and mutual learning,
- Give effective feedback regarding which dominant behaviours should be reinforced or eliminated,
- Invite dominants to be guests in the subordinants group culture,
- Push for constructive change [77].

Members of both subordinant and dominant groups have roles to play in shaping a culture of inclusion. Individuals, however, possess multiple identities, and as such may be members of both dominant and subordinant groups. To assist, therefore, in creating a culture of inclusion in the organization, individuals will likely need to attempt to reconcile their multiple roles and group memberships [77].

Potential Advantages Presented by a Culture of Inclusion

As highlighted in Fig. 2, there are a number of advantages associated with inclusion. Some of these advantages directly benefit employees, while all benefit the organisation. Many of the advantages identified in Fig. 2 mirror those resulting

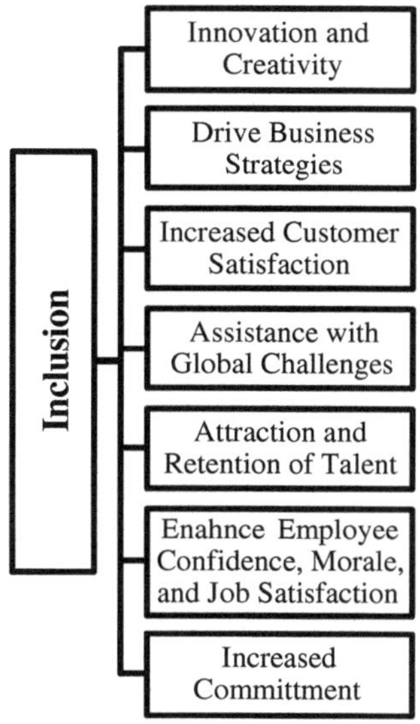

Fig. 2 Advantages associated with inclusion. *Source* Author (O'Donovan 2017)

from diversity in the organization. As discussed earlier, however, the advantages associated with diversity are more likely to be realised under a culture of inclusion. Given that the potential advantages are mirrored, yet inclusion is a more holistic, integrative approach to diversity management, it is again argued that organisations move beyond diversity management towards inclusion. Indeed, a report examining global diversity and inclusion published by Forbes, in 2011, proposes that both a diverse workforce and an inclusive culture are necessary for global success. Similarly, Sabharwal [68] contends that inclusive management holds greater potential for improved productivity and workplace harmony than diversity management alone. Ferdman [79] simply states that inclusion can provide critical benefits to workgroups, organisations and society.

Under an inclusive environment, diversity is a key driver of innovation and creativity, and can drive business strategies [80]. Similar to the business case for managing diversity, diversity coupled with an inclusive culture is purported to enhance idea generation relating to products and practices, due to the wide range of employee experiences and perspectives [65, 80]. Indeed, Davidson and Ferdman [69] have previously asserted that inclusion serves to open a pathway by which individuals can organise and use their personal resources to do what they do best, meaning employees can use those diverse experiences and perspectives. Moreover, 85% of executive respondents in the aforementioned global Forbes [80] report agreed with the assertion that a diverse, inclusive workforce is critical to encouraging the varied ideas and perspectives that drive innovation. Indeed, increased focus may be given by companies to the impact of diversity and inclusion on innovation, as organisations increasingly attempt to use the power of innovation to drive business goals [80]. It is interesting to note that in 2003, during a workshop on the topic of inclusion undertaken by Ferdman, in which participants were asked to discuss outcomes of inclusion that they had experienced, participants noted an improved ability to accomplish organisational goals, again supporting the argument for an inclusionary approach to diversity management.

The existence of a diverse and inclusive workforce can assist in ensuring that the products and services offered by the organisation are respectful and mindful of their customers, and even their customer's clients. If the workforce is reflective of the organisations customer base, and is willing and able to use that similarity to improve product or service offerings, the organisations outputs can be closer aligned with customer desires. This can result in increased customer satisfaction [76, 80]. Inclusive organisations may also benefit from assistance from their employees on global challenges, including varying laws and regulations, language barriers and cultural barriers [80]. Employees in an inclusive organization may, for example be willing to openly discuss their national culture, thereby assisting the organisation in better understanding, and adapting to, that culture. Moreover, inclusion can result in improved productivity and less errors, which can result in the production of products of a higher quality [76]. Indeed, decreased errors may also positively impact the organisations bottom line, as less profit will be lost to wastage.

As discussed earlier in this chapter, the business case for diversity proposes that the employment of diverse employees, and diversity management policies, will aid in the attraction, and retention, of the best talent [14, 17]. Rather than the mere presence of diversity in a workforce, however, an inclusive organisational environment is crucial to the attraction and retention of the best candidates [80]. Organisations can design diversity and inclusion policies to be used as specific recruiting and retention policies, broadening the talent pool from which they can recruit, while developing an employment brand that is seen to be fully inclusive [80]. This nods to a link between inclusion and Employer Branding initiatives, indicating that inclusion can assist in strengthening the employer brand, while the employer brand can be used to reinforce, and highlight, inclusion. Success in recruiting diverse talent does often, however, depend on the organisations ability to provide job advancement opportunities equally. This presents an argument as to why organisations should begin inclusion efforts via diversity management, as one aspect of diversity management discussed earlier was the auditing of policies to ensure fair treatment of all employees.

Inclusion has been identified as carrying the advantage of enhancing employee confidence [1, 76]. Enhancing confidence has the potential to raise an individual's perception of their ability, which can positively impact performance. In addition to enhancing performance, increased confidence may increase the likelihood of the individual believing they are capable of going beyond their required tasks, that is undertaking Organizational Citizenship Behaviours (OCBs), which are supra-role behaviours [1].

Inclusion can also result in enhanced morale, feelings of support and job satisfaction [1]. Interestingly, job satisfaction, engagement and a supportive work environment have been previously cited in the literature as antecedents of OCBs. If inclusion results in these factors, it is therefore again arguably evident that a relationship exists between inclusion and OCBs [1]. It is also arguable that perceiving higher levels of support will increase employee loyalty and commitment to the organisation, thereby increasing the likelihood that they will perform in a manner that exceeds role requirements, perhaps in an effort to reciprocate for the support and increased levels of morale emanating from the organisation.

In a similar vein, a further resultant factor emanating from perceived inclusion concerns increased commitment [1, 76], whether to the job or organisation and enhanced productivity, which are again likely to enhance performance [1]. Additionally, increased commitment has the potential to see reduced absenteeism and turnover levels, allowing for a stability in the workplace, and savings through reduced recruitment and selection costs.

Inclusion can also result in increased knowledge transfer, group cohesion and a more positive group climate, thus, a better work environment [76]. Increased knowledge transfer may pave the way for smoother inter- and cross-departmental functioning, enhancing business operations. Improved group cohesion and a positive group climate will likely assist in the creation of a team culture, which is identified as a factor contributing to perceived inclusion later in this chapter, as well as potentially improving overall group performance.

Lockwood [14] simply summates the above arguments in favour of inclusion by proposing that it serves to generate opportunities for growth, flexibility, and adaption in the marketplace. Sabharwal [68] additionally summates, more generally than above, that inclusion can result in improved organisational performance. Organisations are cautioned, however, that in order to achieve enhanced performance via inclusion, leadership which is both dedicated to fostering inclusion and willing to empower employees in a manner by which they can influence work decisions is necessary. Indeed, Shore et al. [81] also contend that organisations must have leadership dedicated to fostering inclusion at all organisational levels. Furthermore, Davidson and Ferdman [69] deem the development of an organisational culture of inclusion the responsibility of every organisational member, also proposing that if individuals expect inclusion, they must also learn how to provide it. Organisations are cautioned that inclusion efforts are hampered when employees perceive other employees through the lens of oversimplified, negative stereotypes, and when interpersonal interactions are perverted by status dynamics [82].

Consequently, introducing diversity management practices which are specifically targeted at improving the situation of historically disadvantaged groups will likely fail to foster true inclusion [83]. One reason for such an argument is that diversity management initiatives which focus on disadvantaged groups cause resentment on the part of those who do not benefit from the diversity practices, and potentially also exacerbates existing stereotypes [84]. Nishii [83] posits that as organisations increasingly look to innovation to foster long-term success and growth, it is of critical importance that the downside of diversity is addressed. In particular, Nishii [83] suggests that solutions that make productive debate possible, while also enhancing cooperation and learning are required.

An important starting point for such solutions is presented by inclusive climates, which minimise divisive conflict by minimising structural inequalities, assimilation and exclusionary decision-making. Nishii's [83] argument, indeed much of the proceeding section, is supported by Sabharwal [68], who states that Organizational Inclusive Behaviours (that is, behaviours which create inclusion) can be summated under three concepts. Those three concepts are suggested to be commitment from top leadership to foster inclusion, the ability to influence organisational decisions, and fair treatment. Further discussion on fostering or developing inclusion is presented in the following, penultimate section of this chapter.

Inclusion: How to Get There

Stemming from a 2015 study by this chapter author, a number of contributory factors can assist organisations in developing a sense of inclusion, as diagrammatically represented in Fig. 3.

As identified at the beginning of this chapter, it is recommended that organisations begin their inclusion efforts via diversity management initiatives. A tenet of this chapter is that organisations should move beyond traditional diversity management programmes towards an inclusionary approach to diversity. Nonetheless, there is merit in beginning with diversity management. Diversity management facilitates the opening of a dialogue in organisations concerning what diversity is, the potential advantages it affords organisations, and the identification of challenges to workforce diversity. Moreover, via the diversity training elements of diversity management, organisations can encourage employees to consider what makes them different, but not for the purpose of negatively highlighting differences, rather with a focus on how those differences can be harnessed in the workplace. Similarities among individuals who appear quite different may also be focussed upon, thereby increasing understanding and harmony, and enhancing a shared sense of purpose.

Diversity management has the potential, however, to be divisive, regardless of original intent. The process may result in attempting to understand employees by

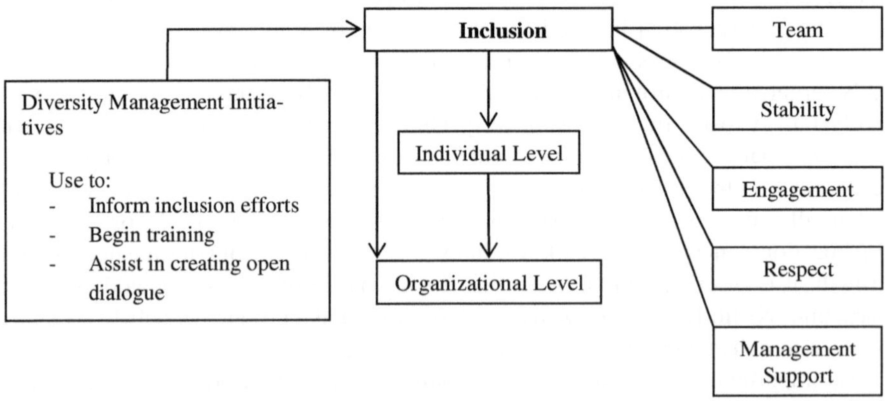

Fig. 3 Factors contributing to inclusion. *Source* Author (O'Donovan 2017)

"assigning" them to, and viewing them in light of assignment to, a particular grouping, for example female, millennial or knowledge worker. This rigid categorisation has the potential to ignore other aspects of an individual's diversity, potentially resulting in individuals feeling marginalised. For example, it does not follow that a "female" identifies as female, nor does it follow that a millennial worker shares common characteristics with fellow millennials. It is recommended, therefore, that while beginning with diversity management initiatives, organisations do so with the ultimate aim of creating inclusion.

As discussed earlier, inclusion concerns holistically viewing the employee, integrating both their similarities and differences into the fabric of the organisation. Essentially, employees are allowed to be their full selves while in the workplace. Moreover, differences and similarities are leveraged to improve organisational functioning and performance. Organisations are advised to take such an approach to diversity, and, more generally, talent management. It was noted earlier that inclusion occurs at both the individual and organisational level, and is a contextual, individual, transient concept. Creating and maintaining inclusion is, therefore, an ongoing process. Regardless, organisations can capitalise on the identification of a number of contributory factors assisting in the perception of inclusion.

The first contributory factor that organisations can develop, encourage and maintain concerns teams. In particular, having a sense of team was strongly identified by O'Donovan [1] as important for inclusion, therefore, organisations should identify ways in which a team orientation can be created. Organisations may, for example schedule team meetings or briefings, or encourage inter- or cross-departmental problem solving.

A second contributory factor concerns stability. In particular, the related concepts of familiarity in a unit or department and the existence of relationships in the workplace, both of which speak to stability, have been established as important. Organisations are therefore advised to assist, or allow, individuals to form and maintain relationships in the workplace. In addition, while rotation has been long proposed as a potential combatant to low motivation or fatigue as a result of monotony, organisations are, for the purpose of developing inclusion, advised to avoid unnecessary, frequent, transfers across departments. If such transfers are necessary, organisations may find it useful to encourage, and facilitate, cross-departmental relationship building, whether by formal or informal means, in an attempt to create familiarity throughout the organization.

Employee engagement also carries significance for inclusion. In particular, perceiving respect from colleagues, in addition to being willing, and believing there exists the freedom, to offer opinions on work-related matters, are further contributory factors. Believing oneself to be respected and free to engage in dialogue in the workplace are elements of employee engagement, again reinforcing the argument in the proceeding section that a link exists between engagement and inclusion. Consequently, organisations should, in conjunction with inclusionary efforts, also focus on employee engagement. Indeed, as concepts, both engagement and inclusion arguably support and reinforce each other.

The final contributory factor identified in Fig. 3 concerns management. In particular, management have a role to play in creating inclusion, much like they do in championing diversity. While what constitutes inclusion is subjective, and therefore seems different to every individual, management have a role to play in creating a sense of inclusion. If managers are seen to behave inclusively, this symbolises the importance of inclusion throughout the organisation. Similarly, if management are seen to visibly encourage individuals to engage their differences to assist in their work, this symbolises acceptance and valuing of differences. More generally, managers have a role to play in reinforcing organisational culture, thus have a role to play in reinforcing a culture of inclusion. Managers can act as driving forces, or champions, for the other contributory factors identified and discussed above, specifically, a sense of team, stability and employee engagement. These factors, whether one, all, a combination, or in conjunction with other individual factors, contribute towards perceived inclusion.

Conclusion

Perceived inclusion results in a number of positive outcomes, which organisations should display an interest in, primarily as these outcomes carry the potential to enhance performance. Inclusion is more likely than the mere presence of diversity, or than diversity management, to result in capitalisation on the benefits associated with diversity. Moreover, inclusion can enhance employee confidence, presenting the potential of raising an individual's perception of their ability, which can positively impact performance and increase the likelihood of the individual believing they are capable of going beyond their required tasks. Inclusion can also result in enhanced morale, feelings of support and job satisfaction, which can improve performance and reduce absenteeism, and speaks to the importance of integrating inclusion and employee engagement initiatives. Further resultant factors emanating from perceived inclusion concern increased commitment, whether to the job or organisation, and enhanced productivity, which are again likely to enhance overall organisational performance.

Finally, it is strongly recommended that organisations develop an organisational culture which truly values, and espouses the benefits of, inclusion in the workforce. This culture can be used as a continual reinforcer of diversity and inclusion efforts. To begin doing so, organisations can make use of the contributory factors identified in this chapter. The factors contributing to perceived inclusion presented in this chapter are far from complicated, making them relatively easily replicable across organisations of different sizes and structures in different industries. Finally, organisations are encouraged to create a culture that enables an open dialogue between employees and managers. The subjective nature of inclusion means that the best placed individuals to tell us what inclusion looks like are the organisation's employees. It is time to move past diversity management, and focus on inclusion; Diversity Management 2.0.

References

1. O'Donovan, D. (2015). *Diversity, inclusion and organizational citizenship behaviours: A study of nurses in the Irish healthcare sector*, Unpublished PhD Thesis, Cork Institute of Technology.
2. Carnevale, A. P. & Stone, S. C. (1994 October). "Diversity: Beyond the Golden Rule", *Training and Development*, 48(10), 22–39.
3. Gordon, J. R. (2002). *Organizational behaviour: A diagnostic approach* (7th ed.). New Jersey: Perason Education.
4. Griffin, R., & Moorhead, G. (2006). *Fundamentals of organizational behaviour: Managing people and organizations*. Boston: Houghton Mifflin.
5. Hite, L. M., & McDonald, K. S. (2006). Diversity training pitfalls and possibilities: An exploration of small and mid-size US organizations. *Human Resource Development International, 9*(3), 365–377.
6. Jackson, S. E., & Joshi, A. (2011). Work team diversity, In Zedeck, S. *APA handbook of industrial and organizational psychology, 1,* 651–686.
7. Monks, K. (2007). *The business impact of equality and diversity: The international evidence*. Dublin: The Equality Authority and National Centre for Partnership & Performance.
8. Morley, M., Moore, S., Heraty, N., Linehan, M., & MacCurtain, S. (2004). *Principles of organizational behaviour: An Irish text* (2nd ed.). Dublin: Gill & Macmillan.
9. Moore, S. (1999). Understanding and managing diversity among groups at work: Key issues for organizational training and development. *Journal of European Industrial Training, 23* (4/5), 208–217.
10. Bechtoldt, M. N., De Dreu, C. K. W., & Nijstad, A. (2007). Team personality diversity, group creativity, and innovativeness in organizational teams. Sustainable Diversity EURODIV Working Paper No.1.2007 [available online] http://www.susdiv.org/uploadfiles/SD2007-001.pdf
11. White, H., & Rice, M. (2010). The multiple dimensions of diversity and culture. *Review of diversity and public administration: Theory, issues and perspectives* (2nd Edn.).
12. Joplin, J. R. W., & Daus, C. S. (1997). Challenges of leading a diverse workforce. *Academy of Management Executive, 11*(3), 32–47.
13. Espinoza, M. (2007). Turning diversity into a competitive advantage. *Financial Executive, 23* (3), 43–46.
14. Lockwood, N. R. (2005). Workplace diversity: Leveraging the power of difference for competitive advantage. *HR Magazine, 50*(6), 1–10.
15. Tipper, J. (2004). How to increase diversity through your recruitment prcatices. *Industrial and Commercial Training, 36*(4), 158–161.
16. Gardenswartz, L., & Rowe, A. (1998). Why diversity matters. *HR Focus, 75*(7), 51–54.
17. Robinson, G., & Dechant, K. (1997). Building a business case for diversity. *Academy of Management Executive, 11*(3), 21–31.
18. Kinicki, A., & Kreitner, R. (2006). *Organizational behaviour: Key concepts, skills & best practices* (2nd ed.). New York: McGraw-Hill/Irwin.
19. Marchington, M., & Wilkinson, A. (2005). *Human resource management at work: People management and develpoment* (3rd ed.). London: CIPD.
20. Von Bergen, C. W., Soper, B., & Foster, T. (2002). Unintended negative effects of diversity management. *Public Personnel Management, 31*(2), 239–251.
21. IBEC. (2003). *Ireland's changing workforce: Harnessing diversity in the workplace*. Dublin: Irish Business and Employers Confederation.
22. Kim, B. Y. (2006). Managing workforce diversity: Developing a learning organization. *Journal of Human Resources in Hospitality and Tourism, 5*(2), 69–90.
23. Cox, T. H., & Blake, S. (1991). Managing cultural diversity: Implications for organizational competitiveness. *Academy of Management Executive, 5*(2), 34–47.

24. Kinicki, A., & Kreitner, R. (2001). *Organizational Behaviour* (5th ed.). New York: McGraw-Hill/Irwin.
25. Carbery, R., & Cross, C. (2013). *Human resource management: A concise introduction.* England: Palgrave Macmillan.
26. Collings, D. G., McDonnell, A., & Scullion, H. (2009). Global talent management: The law of the few. *Poznan University of Economics Review, 9*(2), 5–18.
27. Bagshaw, M. (2004). Is diversity divisive? A positive training approach. *Industrial and Commercial Training, 36*(4), 153–157.
28. Flynn, G. (1995). Do you have the right approach to diversity? Avon, Bank of America and Gannett's models of diversity. *Personnel Journal, 74,* 68–70.
29. Schneider, B. (1987). The people make the place. *Personnel Psychology, 40,* 437–453.
30. Centre for Strategy and Evaluation Service. (2003). *The costs and benefits of diversity: Executive summary*. United Kingdom: Centre for Strategy and Evaluation Services.
31. European Commission, Employment & Social Affairs. (2003). *The costs and benefits of diversity*, United Kingdom: CSES.
32. Farrer, J. (2004). A Practical approach to diversity. *Industrial and Commercial Training, 36* (4), 175–177.
33. Lowther, R. (2006). Embracing and managing diversity at Dell: Introducing flexible working and a women's network to help retain key employees. *Strategic HR Review, 5*(6), 16–19.
34. Anderson, J. A. (1993). Thinking about diversity. *Training And Development, 47*(4), 59–60.
35. Cox, T., Jr. (1991). The multicultural organization. *Academy of Management Executive, 5*(2), 34–47.
36. Waters, H., Jr. (1992). Minority leadership problems. *Journal of Education for Business, 68* (1), 15–20.
37. Eisenberger, R., Fasolo, P. M., & Davis-LaMastro, V. (1990). Effects of perceived organizational support on employee diligence, innovation, and commitment. *Journal of Applied Psychology, 53,* 51–59.
38. van Knippenberg, D. L. (2007). "Understanding diversity", *Inaugural address series: Research in management*. Roterdam: Erasmus Research Institute of Management.
39. Justesen, S. L. (2001), *A study of the dynamics inherent in the relationahip between innovation and diversity*, Master's Thesis, Denmark: Copenhagen Business School, cited in Bechtoldt, M.N., De Dreu, C.K.W., Nijstad, A. (2007), "Team Personality Diversity, Group Creativity, and Innovativeness in Organizational Teams", Sustainable Diversity EURODIV Working Paper No.1.2007 [available online] http://www.susdiv.org/uploadfiles/SD2007-001.pdf
40. Crockett, J. (1999). Winning competitive advantage through a diverse workforce. *HR Focus, 76*(5), 9–10.
41. Doke, D., & Beagrie, S. (2003). How to…Manage a diversity programme. *Personnel Today, 2*(4), 31–34.
42. Ferguson, J. T., & Johnston, W. R. (1995). Managing diversity. *Mortgage Banking, 55*(12), 33–36.
43. Miller, S. K., & Tucker, J. J. (2013). Diversity trends, practices, and challenges in the financial services industry. *Journal of Financial Service Professionals,* 46–57.
44. Kreitz, P. A. (2008). Best practices for managing organizational diversity. *The Journal of Academic Librarianship, 34*(2), 101–120.
45. United States Government Accountability Office. (2005). *Diversity management: Expert-identified leading practices and agency examples*, Washington DC: GAO, [online], available: http://www.gao.gov/new.items/d0590.pdf, Accessed September 14, 2016.
46. Liff, S. (1997). Two routes to managing diversity: Individual differences or social group characteristics. *Employee Relations, 19*(1), 11–26.
47. Miller, E. K. (1994). Diversity and its management: Training managers for cultural competence within the organization. *Management Quarterly, 35*(2), 17–23.

48. Flynn, G. (1999). White males see diversiy's other side. *Workforce, 78*(2), 52–55.
49. Daas, P., & Parker, B. (1999). Strategies for managing human resource diversity: From resistance to learning. *Academy of Management Executive, 13*(2), 68–80.
50. Muzio, D., & Tomlinson, J. (2012). Editorial: Researching gender, inclusion and diversity in contemporary professions and professional organizations. *Gender, Work and Organization, 19*(5), 455–466.
51. Smithson, J., & Stoke, E. (2005). Discourses of work-life balance: Negotiating @gender-blinded@ terms in organizations. *Gender, Work and Organization, 12*(2), 147–168.
52. Kent, P. & O'Donovan, D. (2016). What is the impact of flexible working on the worker and the workplace? Paper presented at the *Irish Academy of Management* Conference Proceedings, University College Dublin, August 31st to September 2nd.
53. Marina, B. (2010). Managing diversity in communities, workplaces and society, *Review of diversity and public administration: Theory, issues and perspectives* (2nd Edn.).
54. Mannix, E., & Neale, M. A. (2005). What differences make a difference: The promise and reality of diverse teams in organizations? *Psychological Science in the Public Interest, 6*(2), 31–55.
55. Thomas, D. A., & Ely, R. J. (1996). Making differences matter: A new paradigm for managing diversity. *Harvard Business Review, 74*(5), 79–90.
56. Lussier, R. N. (2008). *Human relations in organizations: Applications and skill building* (7th ed.). New York: McGraw-Hill/Irwin.
57. O'Donovan, D., & Linehan, M. (2010). Implications of cultural diversity for performance management systems in Irish hotels. Paper presented at the Irish Academy of Management Conference Proceedings, Cork Institute of Technology, Cork, September 1st–3rd.
58. Robbins, S. P. (2003). *Organizational behaviour: International edition* (10th ed.). New Jersey: Prentice Hall.
59. Ferdman, B. M. (2013). Frameworks for Understanding Inclusion. In B. M. Ferdman, & B. R. Deane (Eds.) *Diversity at work: The practice of inclusion*, Jossey-Bass.
60. Friday, E., & Friday, S. W. (2003). Managing diversity using a strategic planned change approach. *Journal of Management Development, 22*(10), 863–880.
61. Thomas, R. R., Jr. (1990). From affirmative action to affirming diversity. *Harvard Business Review, 68*(2), 107–117.
62. Popescu, S., & Rusko, R. (2012). Managing diversity in public organizations. *Global business and management research: An international journal, 4*(3–4), 235–247.
63. Jones, R. T., Jerich, B., Copeland, L., & Boyles, M. (1989). Four by four: How do you manage a diverse workforce? *Training and Development Journal, 43*(2), 13–21.
64. Kochan, T., Bezrukova, K., Ely, R., Jackson, S., Joshi, A., & Jehn, K. (2003). The effects of diversity on business performance: Report of the diversity research network. *Human Resource Management, 42,* 3–21.
65. Ferdman, B. M., & Davidson, M. N. (2002). A matter of difference—Drawing the line: Are some differences too different? (Or: Who's in, who's out, and what difference does it make?). *The Industrial-Organizational Psychologist, 39*(3), 43–46.
66. Ferdman, B. M., & Brody, S. E. (1996). Models of diversity training. In D. Landis & R. S. Bhagat (Eds.), *Handbook of intercultural training* (2nd ed.). Thousand Oaks, CA: Sage.
67. Pless, N. M., & Maak, T. (2004). Building an inclusive diversity culture: Principles, processes and practices. *Journal of Business Ethics, 54*(2), 129–147.
68. Sabharwal, M. (2014). Is diversity management sufficient? Organizational inclusion to further performance. *Public Personnel Management, 43*(2), 197–217.
69. Davidson, M. N., & Ferdman, B. M. (2002). Inclusion: What can I and my organization do about it? *The Industrial-Organizational Psychologist, 39*(4), 80–85.
70. Ferdman, B. M. (1995). Cultural identity and diversity in organizations: Bridging the gap between group differences and individual uniqueness. In M. M. Chemers & S. Oskamp (Eds.), *Diversity in organizations: New perspectives for a changing workplace* (pp. 37–61). Thousand Oaks, CA: Sage.

71. Nishii, L., Rich, R., & Woods, S. (2006). *Framework of organizational inclusion*. New York: Cornell University ILR.
72. Holvino, E., Ferdman, B. M., & Merrill-Sands, D. (2004). Creating and sustaining diversity and inclusion in organizations: Strategies and approaches. In M. S. Stockdale, & F. J. Crosby (Eds.) *The psychology and management of workplace diversity*, 245–276.
73. Andresen, M. (2007). Diversity learning, knowledge diversity and inclusion: Theory and practice as exemplified by corporate universities. *Equal Opportunities International, 26*(8), 743–760.
74. Davidson, M. N., & Ferdman, B. M. (2001). Diversity and inclusion: What difference does it make? *The Industrial-Organizational Psychologist, 39*(2), 36–38.
75. Gasoreck, D. (2000). Inclusion at Dun & Bradstreet: Building a high-performing company. *The Diversity Factor*, Summer, 25–29.
76. Ferdman, B. M. (2003). Accounts of inclusion (and exclusion). *The Industrial-Organizational Psychologist, 40*(4), 81–86.
77. Davidson, M. N., & Ferdman, B. M. (2002). Inclusion and power: Reflections on dominance and subordination in organizations. *The Industrial-Organizational Psychologist, 40*(1), 62–67.
78. Wishik, H. R., & Davidson, M. N. (2002). Learning to lead: Some impacts of social identity on management and leadership development. In M. N. Davidson, & B. M. Ferdman, Inclusion and power: Reflections on dominance and subordination in organizations. *The Industrial-Organizational Psychologist, 40*(1), 62–67.
79. Ferdman, B. M. (2007). The slippery slope of inclusion. *San Diego Psychologist, 22*(7), 13–18.
80. Forbes. (2011). *Global diversity and inclusion: Fostering innovation through a diverse workforce*. New York: Forbes Insights.
81. Shore, L. M., Randel, A. E., Chung, B. G., Dean, M. A., Ehrhart, K. H., & Singh, G. (2011). Inclusion and diversity in work groups: A review and model for future research. *Journal of Management, 37,* 1262–1289.
82. DiTomaso, N., Post, C., & Parks-Yancy, R. (2007). Workfoce diversity and inequality: Powerr, status, and numbers. In D. S. Massey, & K. S. Cook (Eds.), *Annual Review of Psychology, 33*, 473–501.
83. Nishii, L. H. (2013). The benefits of climate for inclusion for gender-diverse groups. *Academy of Management Journal, 56*(6), 1754–1774.
84. Fiol, C. M., Pratt, M. G., & O'Connor, E. J. (2009). Managing intractable identity conflicts. *Academy of Management Review, 34,* 32–55.

Hard Times, Less Compassion? Distinct Perspectives Towards Distinct Minorities in the Portuguese Organizational Context

Iris Barbosa

Abstract Several approaches to manage diversity were developed in the U.S. and then disseminated to Europe. Their origin can be drawn in a timeline, but not their end. Research is still needed on the way organizations combine them towards distinct minority groups, particularly in less explored national contexts. The present qualitative study examines how minorities are evaluated and dealt with by large organizations operating in Portugal. The results reveal that distinct minorities are being approached distinctively and that this significantly stems from the country's current hard financial conditions and ensuing social challenges, as well as from the colonialist Portuguese past. The study particularly contributes by exposing diversity as a social construct that can assume as many facets as the minority identity groups present in a specific space and time.

Introduction

Recent studies revealed that affiliates from foreign origin are contributing to introduce the diversity rhetoric in Portugal [1], although sometimes with trivial practical impact [2]. Efforts to translate that discourse to the Portuguese context were also reported, consisting mostly of its combination with the more traditional equality/nondiscrimination principles and/or the integration of the subject in the Corporate Social Responsibility (CSR) field [2]. The case of women as a political minority and their struggle to reconcile increasingly demanding jobs with the domestic and familiar responsibilities has been especially analyzed by Portuguese academics [3–5]. However, insufficient attention has been paid to the way distinct minority groups are being evaluated and dealt with in the current context of scarce resources and job opportunities in Portugal. This is a particularly relevant scenario to develop a study of this nature since at the same time that (a) it is expected to

I. Barbosa (✉)
Department of Management, School of Economics and Management,
University of Minho, Campus Gualtar, 4710-057 Braga, Portugal
e-mail: iris@eeg.uminho.pt

C. Machado and J.P. Davim (eds.), *Managing Organizational Diversity*,
DOI 10.1007/978-3-319-54925-5_2

exacerbate identity attributes and thus fuel prejudice and discrimination [6], (b) a diverse workforce that fosters innovation and competently deals with diverse markets [7] seems vital.

Hence, the present study addresses the following research questions: How do organizations operating in Portugal evaluate the presence of specific minority groups? What organizational policies and practices are being developed and applied to specific minority groups? How do they reflect the distinct diversity approaches originated in the US, such as defined on Dass & Parker's [8] typology? Aiming to get an overview of the subject, a content analysis was made of the 24 corporate websites (out of the group of the 500 largest companies operating in Portugal) that presented diversity related contents in Portuguese language. Subsequently, an exploratory research was conducted on six of those companies, which relied on interviews with both managerial and nonmanagerial members and on the analysis of relevant corporate documents. The results highlight that *diversity* is not a monolithic concept. Instead, distinct identity groups can be evaluated and dealt with in specific ways in the same organization and at the same time. In particular, organizations may, consciously or not, avoid, tolerate, or appreciate the presence of distinct identity groups. The study also contributes by revealing that both the colonialist Portuguese past and the current financial crisis in Portugal have an impact on how minorities are being perceived and managed at the workplace.

Conceptual Background

Minorities *and Other Related Concepts*

The term *minorities* usually refers to groups of people that, in a particular social context, mainly national, are less common or do not characterize the majority of the population (i.e. *the dominant group*), but *the others* [6, 9]. Some minority identity attributes are related to the notion of disadvantage, like psychical and mental disabilities and chronic diseases [10]. Other human particularities, such as specific ethnicities and religions, may also be associated with character or competency flaws, as well as seen of less legitimacy in a given social/national context [11]. Hence, social and organizational prejudice (negative feelings) and discrimination (detrimental behaviors) are common outcomes for minority groups [6]. Despite their numeric representation, women constitute a political minority, since they are frequently considered of inferior status than men [12] and underrepresented in high-paying jobs and top management functions [13].

Several scholars proposed classifying the vast set of identity attributes mostly by discerning the more visible human traits from the less visible or intrinsic ones [14–17]. The more visible differences include biologic features such as gender, skin colour, age group, and notorious physical disabilities. The opposing side of the catalogue presents less detectable features (e.g. nationality of origin, ethnicity, sexual orientation, and chronic disease), which individuals may opt to pass or camouflage when perceiving less tolerant societies and/or organizations [14, 18]. *Difference* is thus conceived as a social construct bind to space and time. Change is always possible, as exemplified by the increasing social acceptance towards minority lifestyles in big urban centres. But renewed intolerance towards *the outsider* can also occur, such as expected in societies facing severe financial conditions and few job opportunities [6].

Main Approaches to Deal with Workforce Diversity

The remarkable human heterogeneity that weaves the U.S. social fabric early impelled this country to find effective ways to deal with diversity at both social and organizational levels. The European countries have been significantly influenced by most of such proposals, which encompass particular views towards minorities. In the early twentieth century, the U.S. *melting-pot* ambition implied that minorities were accepted as long as they renounced to their cultural identity. By the 1950's–60's, the increasing proportion of ethnic minorities and their refusal and/or inability to blend in led the established majority to perceive them as a threat to the status quo [19]. The assimilation ideal was then replaced by what Dass and Parker [8] refer as the *resistance perspective* since this approach denies or avoids the difference, which also characterized the European post-colonial period.

Originated in 1964, the U.S. Equal Opportunities legislation recognizes the social groups that were historically discriminated against and demands organizations to adopt individual merit as the unique criterion for human resource decisions [20]. This legal approach is deemed fundamental to circumvent biased decisions [21], but falls short by ignoring the structural disadvantages of minorities. Moreover, it allows organizations to limit the presence of certain minority groups by claiming (apparent) meritocracy reasons, such as non-appropriate education or work inexperience. Equality/non-discrimination principles and laws integrate the national constitutions of the European countries, as well as the European Community legislation [22].

The recognition that the equality legislation was not sufficient to overcome the effects of past discrimination fueled the development of the U.S. Affirmative Action policies in the same decade [23]. Accordingly, organizations should promote the presence of historically disadvantaged groups, such as women and ethnic minorities. This approach was modestly implemented across Europe and its coercive character is usually dismissed. Member states are recommended to design training programmes to overcome the structural and educational barriers faced by ethnic minorities and older people, as well as to promote the integration of people with physical disabilities. Dass and Parker [8] classify the Equal Opportunities and the Affirmative Action policies as *Discrimination and Fairness* type of approaches, which are particularly characterized by conceiving differences as the cause of problems and, as a result, by demanding minorities' assimilation into the established culture.

The 1980's context of growing global competition assisted to the development of a strategic and voluntary approach that perceives differences as *creating opportunities*, such as defined by Dass and Parker [8]. In special, ethnic minorities are believed to help organizations respond to the needs of customers from the same cultural backgrounds [24]. Moreover, the exhibition of a diverse workforce is deemed to improve the organizational image among stakeholders who increasingly monitor companies' ethics and social responsibility initiatives [29]. Dass and Parker [8] evaluate this perspective as *legitimizing the access* of minorities to better jobs, but other authors criticize it on the grounds of its *essentialist nature* [16]—i.e. minorities are seen as serving particular business interests, but the individual's unique vocation and skills may well be ignored. The discourse that minorities bring important business advantages is frequently considered to be at odds with the traditional European values of equality and meritocracy [25].

The U.S. diversity management model, or multicultural approach, conceives diversity as a broader concept by including all human differences, and not just the ones traditionally protected by the legislation. According to Dass and Parker [8], it also assumes that individuals from distinct identity groups share similarities and that both difference and sameness must be managed in the interest of long-term *learning and effectiveness*. A diverse workforce is seen as offering the organization a wide spectrum of critical analyzes and innovative ideas [7]. Opponents claim, though, that the assumption that everybody is unique masks the structural disadvantages of specific minority groups [26]. The U.S. Diversity Management discourse seems to have conquered agenda among the E.C. politics. The European Commission website, link Tackling Discrimination, reads "The *Business Case for Diversity* shows that diversity management—whereby employers recognize, value and include women and men of different ages, abilities, ethnic origin, religion or sexual orientation—makes good business sense" [27]. However, investigations indicate that European companies usually translate the diversity model by mixing it with the more traditional equality rhetoric and social responsibility goals [28–31], thus revealing some resistance to embracing it.

Managing the Portuguese Social Diversity

From the fifteenth century to the last third of the twentieth century, Portugal carried out a significant colonialist venture. Native individuals from the several Portuguese colonies (e.g. Angola, Mozambique, and Brazil) were seen and treated as intellectually and morally inferior [32], an approach that persisted after the slavery period. Consequently, these minorities were typically confined to the less rewarding functions, usually more physically demanding, and expected to observe the Portuguese cultural patterns. With the end of the dictatorial regime in 1974 and consequent loss of the last colonies in Africa, more than half a million of the Portuguese people that had settled there came to mainland Portugal. This massive influx of people put extra pressure on a country that was the poorest nation in

Europe at the time [17], with extremely vulnerable finances and limited job and house markets. At least one decade was needed until these individuals were considered properly integrated into the established Portuguese society and seen as regular citizens. The integration into the E.C. in 1986 offered social, political, and economic stability to Portugal, attracting immigrants, mainly from the old African colonies and Brazil, as well as the Eastern Europe. The Portuguese policies of moderate tax and low salaries appealed to international investors and companies [33, 34], which contributed both to introduce foreign management knowledge in Portugal and increase its social diversity through the presence of expatriates.

However, the twenty-first century brought particularly tough challenges to Portugal. A number of local organizations closed or downsized and several international companies relocated their facilities to more attractive economies [35]. High unemployment rates and fiscal pressures, accompanied by a persistent contraction on social protection, culminated in a social crisis. Hard financial sacrifices are now being imposed to most of the population, in a remarkably different situation from two decades ago, when more people enjoyed a medium class existence. Not surprisingly, the 2014 OECD report indicated that the Portuguese people revealed one of the lowest levels of job satisfaction among member countries. Concurrently, social and workforce diversity in Portugal is still increasing. In a country where one income hardly supports a family, more women and older people demand huge participation and opportunities at work. The once omnipresent control of the Catholic Church over the Portuguese society does not seem so relevant nowadays and the definition of *proper* behaviors and lifestyles admits a wider spectrum, particularly in the major cities. On the other hand, the significant number of immigrants in Portugal originated noteworthy religious diversity, especially in the urban area of Greater Lisbon [36]. Moreover, great changes in the Portuguese legislation, namely the ones legalizing abortion and the marriage and adoption of children by same-sex couples are expected to nourish a more openly diverse society and contribute to diminish the reported negative feelings and discomfort of Portuguese people towards homosexuality [37].

In line with the increasing social diversity in Portugal, affiliates from foreign MNCs are contributing to introduce the U.S. diversity rhetoric into the Portuguese organizational landscape [1], although more progressive diversity policies and practices are still dismissed [2]. Yet, much less is known regarding how organizations (and their distinct actors) in Portugal evaluate, portray, and deal with specific minority groups in a context that combines increasing social heterogeneity and diversity rhetoric with scarce resources and job opportunities.

Methods

Previous research has found that only 54 websites from the 500 largest companies operating in Portugal (either from national or foreign origin/ownership control) displayed workplace diversity related contents [1]. Stage 1 of the current study analyzes how the 24 corporate websites of that group that display such contents in Portuguese language portray and declare to manage distinct minority identity groups. Table 1 identifies these companies' main features, 11 of them being from national origin/ownership control and 13 from foreign origin/ownership control. Websites only presented in foreign language were dismissed on the assumption that they do not specifically address the Portuguese national context (but probably a global/international audience). It is worth noticing that a previous comparison was made between each of the 24 websites in Portuguese language and its corresponding English version (whenever the latter existed). Most of the native companies display similar equality/diversity contents in their local and international versions, but companies from foreign origin frequently reveal less developed and/or progressive contents in their Portuguese websites. This result adds to the conviction that the websites' contents in Portuguese language are the ones that address this specific national context, thus offering a more accurate picture on how distinct minority groups are perceived and dealt with in the corresponding company in Portugal.

Table 1 Main features of the 24 companies whose websites were analyzed

Designation	Sector of activity	URL in 2010
Companies from national origin/ownership control (also referred as *native companies*)		
CIRES—Companhia Industrial de Resinas Sintéticas, S.A.	Chemistry industry	www.cires.pt
Companhia Portuguesa de Hipermercados, S.A.	Food distribution	www.auchan.pt
Cooprofar—Cooperativa dos Proprietários de Farmácia, C.R.L.	Pharmacy distribution	www.cooprofar.pt
EDP Distribuição—Energia Lisboa (LVT)	Electricity distribution	www.edp.pt
Lactogal—Produtos Alimentares, S.A.	Agro-industry	www.lactogal.pt
Novadelta—Comércio e Indústria de Cafés, S.A.	Agro-industry	www.delta-cafes.pt
Petrogal Lisboa (LVT)	Fuel distribution	www.galpenergia.com
Recheio–Cash & Carry, S.A.	Food distribution	www.jeronimo-martins.pt
Salvador Caetano—Indústrias Metalúrgicas e Veículos Transporte, S.A.	Metallurgy industry	www. gruposalvadorcaetano.pt
Sumol—Gestão de Marcas, S.A.	Beverage industry	www.sumolis.pt
Unicer—Distribuição de Bebidas, S.A.	Beverage industry	www.unicer.pt
Companies from foreign origin/ownership control (also referred as *foreign companies*)		
BP Portugal Porto Salvo (LVT) (U.K.)	Fuel distribution	www.bp.pt
Companhia IBM Portuguesa, S.A. (U.S.)	IT services	www-5.ibm.com/pt
Eurest Portugal—Sociedade Europeia de Restaurantes, Lda. (U.K.)	Restaurant and catering industry	www.eurest.pt

(continued)

Table 1 (continued)

Designation	Sector of activity	URL in 2010
Glaxosmithkline—Produtos Farmacêuticos, Lda. (U.K.)	Pharmaceutical industry	www.gsk.pt
Gespost—Gestão e Administração de Postos de Abastecimento, Lda. (Spain)	Fuel distribution	www.repsolypf.com
Mapfre—Seguros Gerais, S.A. (Spain)	Insurances	www.mapfre.pt
MSFT—Software para Microcomputadores, Lda. (U.K.)	Business software & services	www.microsoft.com/Portugal
Robert Bosch, Unipessoal, Lda. (Germany)	Automotive and industrial technology	www.bosch.pt
Siemens, S.A. (Germany)	Electrical and electronics industry	www.siemens.pt
Sociedade Portuguesa do Ar Líquido Arliquido, Lda. (France)	Chemistry industry	www.airliquide.pt
Tabaqueira, S.A. (U.S.)	Agro-industry	www.philipmorrisinternational.com/PT
Vodafone Portugal—Comunicações Pessoais, S.A. (U.K.)	Telecommunications services	www.vodafone.pt
Xerox Portugal—equipamentos de Escritório, Lda. (U.S.)	Electrical and electronics distribution	www.xerox.com

All 24 companies were invited to participate in stage 2 of the study, which aimed to further investigate the corresponding diversity perspectives and practices, but only six agreed to. Fictional names are used to protect their anonymity. N1 and N2 are *Native* (N) companies, while F3, F4, F5, and F6 are affiliates from *Foreign* (F) multinational corporations. A total of 44 members of these companies were interviewed, including the Human Resources directors, managers from distinct levels and areas, and collaborators with nonmanagerial positions (workers). The sample reveals diverse in gender (18 men, 26 women) and age (from 22 to 60). Most of the individuals were native Portuguese and five were foreigners from Angola, Mexico, Mozambique, and Zimbabwe. Interviews lasted 40–180 min and 43 of them were audio-recorded with previous consent. Individuals shared their perceptions on how their organizations access and manage specific minority groups, and some added personal experiences. Corporate documents from the six companies (e.g. websites, ethical codes, mission and values statements, corporate magazines) were examined in pursuit of references to diversity and, particularly, minority groups. The companies' demographics were also considered. NVivo 8 supported the analysis of both interviews and the documents considering the analytical categories and subcategories described on Table 2.

Table 2 Analytical categories and subcategories

Analytical categories	Subcategories
Generic impressions towards equality and diversity	• Relevance of a diverse workforce • Relevance of diversity initiatives • Workplace discrimination and prejudice
Diversity dimensions/minorities represented	• Gender and parenthood • Nationality of origin/ethnicity • Age • Disability • Look and appearance • Religion and political ideology • Sexual/affective orientation
Jobs occupied by members of specific minority groups	• Function • Hierarchical level • Associated responsibilities and rewards
Benefits associated with the presence of specific minority groups	• Enhanced performance • More creativity and potential to innovate • Improved decision-making processes • Better response to diverse/international markets • Enhanced corporate image and reputation • Response to institutional expectations • Better work environment • Increased flexibility • Occupation of less interesting and rewarding functions • Employees with less negotiation power
Disadvantages associated with the presence of specific minority groups	• Integration problems • Performance problems • Interpersonal conflicts
Policies and practices directed to specific minority groups	• Identification of the implemented policies and practices (e.g. minorities' recruitment, integration, training and career development, diversity training, diversity councils, work-life balance initiatives) • Corresponding motivations, aims and impact • Perception of corresponding relevance/adequacy
Organizational prejudice/discrimination towards specific minority groups	• Manifestations • Consequences

Dass and Parker's [8] typology of the several diversity perspectives (described on Section "Main Approaches to Deal with Workforce Diversity") was developed considering the U.S. context. However, it seems to be an appropriate tool to assist in the interpretation and categorization of the current study's results given that most of those diversity perspectives were also identified in the European context. Moreover, the coexistence of several approaches is a possibility since this study aims to investigate the views and practices of companies from varied national origins regarding distinct minority groups. The use of these authors' comprehensive typology is thus expected to be of particularly assistance to such analysis.

Distinct Minorities, Distinct Perspectives

Most of the corporate websites from national companies specify the identity attributes they conceive as particularly susceptible to suffer from prejudice and discrimination, as exemplified next:

> Equal conditions for all applicants are guaranteed in the HR recruitment processes, regardless of race, nationality, gender, religion, colour, marital status, social class, and political and sexual orientation.

Websites from companies from foreign origin/ownership control usually conceive diversity as a more inclusive concept that can bring advantages to the business. This is particularly noticeable in the case of companies from U.S. and U.K. origin. The next example is illustrative:

> The company nurtures its cultural roots and interprets diversity as a value and condition for success. Openness and tolerance build a work environment that fosters each member's creativity, innovation and dynamism.

The next sections of this chapter analyze how distinct minority groups are portrayed in the 24 analyzed websites, as well as evaluated and dealt with by the six companies participating in stage 2 of the study.

Gender

Whenever the websites' diversity discourse refers specific identity attributes, gender is always addressed. However, while native companies tend to merely assure the existence of equal opportunities between men and women, several foreign companies express their commitment to promote gender balance. The next quotes are illustrative:

> The same selection conditions are guaranteed for all candidates, regardless of gender, race or religion.

> We are working to increase the presence of women on the board of directors. To achieve it, we are improving our promotion and recruitment practices and rethinking our assessment methods.

N1 managers positively evaluated a gender balanced workforce, considering that "it provides a healthier work environment" (Portuguese male middle manager). However, this view has not stimulated concrete initiatives on the part of this company, such as work-life balance measures. Parenthood, for instance, is faced as a private matter that only concerns the individual, as a female middle manager with a baby explained:

> Children are time and energy sponges and sometimes it is hard to keep the former work standards [before parenthood]. But if you are truly committed, somehow you will work it out.

N2 managers declared to value "women's soft skills, like interpersonal ones" (Portuguese female top manager) as well as the contribution of female workers to "a healthier work environment in the assembly lines" (Portuguese female line manager). This company has developed substantial work-life balance initiatives (e.g. the 67 free hours/year policy for child care) that, according to the HR director (Portuguese male) aim to "alleviate the collaborators' family and household responsibilities, particularly women's, but also to promote women's career development".

F3 reveals a remarkable proportion of women (85%) in the production units, as well as their significant presence in administrative functions. However, men clearly dominate this company's management positions. No formal work-life balance policies and practices were ever defined, but some schedule flexibility is permitted in the administrative and management functions. In contrast, the HR director (Portuguese female) explained that such measures "are impossible to implement in the production units, since work is organized in teams and has to follow a strict schedule" (Portuguese female HR director).

F4 website proclaims that "[F4] promotes and supports a diverse workforce", but this discourse did not prevent this organization from revealing significant gender imbalance: only 25% of the workforce is female and merely 22% of the management positions are occupied by women. Acknowledging this problem, the company presents gender balance as its major diversity goal and is willing to hire and promote more women. A generous set of work-life balance initiatives (e.g. flexible work schedule and a number of convenience services such as babysitting and takeaway) was also developed to help reduce gender imbalance, as concluded by the HR director (Portuguese female): "To promote the presence of women is our main diversity goal. (…) Our extensive list of work-life balance measures assists in this goal".

Women are well represented in F5's management, administrative and R&D functions. However, their access is denied when it comes to operational functions, a policy that is explained and perpetuated by the long lasting lack of feminine dressing rooms. Male workers naturalize gender segregation: "Everybody knows that some functions are for men, particularly if they are physically demanding" (Portuguese male worker). Apparently, the North-American HQs did not stimulate the affiliate to address this problem or to develop work-life balance practices.

At F6, the predominance of women in assembly line posts and men at the engineering functions is argued to result from "the profiles recurrently exhibited by the corresponding applicants, not from any organizational policy or practice" (Portuguese female HR director). Moreover, the interviewed managers contemplate that great gender balance on management positions is particularly promoted by the company's significant work-life balance initiatives (e.g. the free summer activities for the collaborators' children). However, this topic does not seem to be captivating great attention among workers, who are more particularly concerned with job stability. The next quote exposes this perspective: "Frankly, I don't care for gender equality. What I care most about is keeping my job. I don't feel very demanding nowadays" (F6, Portuguese female assembly line worker).

Nationality of Origin/Ethnicity

In the websites from both national and foreign companies, pictures representing collaborators and clients/consumers usually reveal men and women with the kind of appearance that people from Portuguese origin associate with themselves (particularly, brown hair and medium-light skin colour). On the other hand, pictures of people from ethnic minorities tend to figure on the CSR sections/links. The few foreign individuals employed in N1 are from Brazil, Eastern Europe, and African ex-colonies. Managers considered them as "neither an obstacle nor a benefit to the company" (Portuguese female HR director). They occupy less demanding and rewarding operational functions, particularly in the assembly lines, and their presence has not stimulated any specific socialization and integration measures. Moreover, Portuguese native workers revealed little sympathy for the difficulties faced by ethnical minorities, considering that "this is a time of great difficulties for everybody, not just for [ethnic] minorities!" (Portuguese female worker).

N2 facilities in Morocco employ a significant number of native people in both operational and management functions. They are perceived as contributing to the company's international aims since "better than us [the standard 'Portuguese workers'], they know the culture, needs and tastes of clients from the same culture" (Portuguese male middle manager). However, this business case argument, often deemed as "essentialist" [16], has not pervaded the official N2 discourse, since "the Portuguese audience would find it odd" (Portuguese male top manager). In the Portuguese facilities, some foreign individuals from more developed European countries (e.g. England and Denmark) integrate management and R&D functions. However, immigrants from Africa and Eastern Europe are confined to assembly line posts, a situation that the HRM Director (Portuguese male) justifies with "the corresponding inferior education and type of work experience".

F3 employs a remarkable mix of people from 17 less developed countries in their production units, mainly from Eastern Europe and ex-African colonies. Managers assess their presence as being an asset to the company, given their minor status, ambition, and negotiation power: "These people are grateful for the opportunities they have found here, so they are more willing to make sacrifices than Portuguese employees, such as working extra hours and on weekends" (Portuguese male line manager). Also in this company, the few foreigners occupying management functions are from more developed countries, such as the UK. Some Portuguese native workers considered that the development of policies and practices promoting the presence of ethnical minorities was inappropriate or even unfair, since "They [ethnic minorities] are absorbing the few available jobs and that's not fair!" (F3 Portuguese male worker).

F4 has a significant number of collaborators from foreign origin, such as Belgian, Mexican, Angolan, and Cape Verdean. The presence of these individuals is valued and promoted since they are perceived as "offering cultural and linguistic skills that help the company in better communicating and fulfilling the needs of diverse clients" (Portuguese female HR director). Two dark-skinned collaborators from African origin described experiences of discrimination and prejudice in the Portuguese society, but categorically denied similar treatment at F4.

At F5, there is one Belgian top manager and one British R&D employee, but the few employed Brazilian and East European individuals occupy less demanding and rewarding functions at the production lines. A Portuguese male middle manager clearly associates the individual's country of origin with the opportunities found in this company:

> We [at the affiliate] are not unlike the Portuguese society and culture. So, we support all these defeating segregation and racism principles, but when confronted with a real situation, our attitude may be quite different. For instance, we may dismiss a candidate from an ethnic minority arguing that he/she does not have the proper education or experience, but, in fact, we're camouflaging prejudice and stereotypes. (…) A non-white person who comes from a less developed country (such as an old African colony) is more accepted in a lower level task, no doubt. A French, English or German person easily integrates a top position because our mindset associates them with higher knowledge and competence.

F6 employs a significant number of foreign individuals (mostly technicians from Japan, Ireland, England and Germany), since the "company values their singular skills and experiences" (Portuguese female HR director). The organization promotes their smooth integration in both the company and the Portuguese society (e.g. helping them to find suitable houses and schools for their children). Curiously, though, there are no individuals from Africa, Brazil, and/or Eastern Europe, which represent the major part of the Portuguese immigration pool.

Age

Websites from national companies usually display pictures of younger male and female adults, while the pictures presented by foreign companies' websites tend to reveal higher age diversity for both genders. At N1, age diversity was positively evaluated by managers, who find that "it's important to mix the dynamic youth with the responsibility and wisdom of the elder ones" (Portuguese male line manager). On the contrary, the preference for a younger workforce was detected in both N2 and F3, where negative serotypes prevail:

> It's a matter of fact that older people are more resistant to change and less willing to embrace challenges. (N2, Portuguese male HR director)

> Older people lack the strength and dynamism to deal with the hard work in the production units. (F3, Portuguese male manager)

Managers from F4, F5, and F6 did not reveal a preference towards either younger applicants/workers or more experienced ones, and no contradictory evidence was found in these companies' demographics.

Disability

Among the analyzed 24 websites, only one from a native company and three from foreign companies have specific diversity/equality sections. In such cases, pictures representing collaborators with notorious physical disabilities (mostly in wheelchairs) are displayed. N1's non-discrimination discourse, presented in both the website and the lobby's chart values, alludes to the condition of disability. Despite this rhetoric, no individual with this condition works at N1 as managers consider that "it would be too complicate to hire someone with severe disabilities, such as a person in a wheelchair" (Portuguese male middle manager). This opinion and practice seems unjustified both in terms of architectural barriers (since the new facilities have elevators and ramps) and the available functions (for instance, the administrative functions and call centre posts available could be occupied by people with disabilities).

N2 managers depict the integration of people with disabilities as a relevant social responsibility aim. However, individuals with this profile are kept apart from external stakeholders, as a top manager (Portuguese female) explained: "Let's be honest, clients and visitors don't want to come face to face with workers with disabilities. Other functions, where they are not so exposed, and don't feel so observed, are better for them".

F3 managers presented the demanding coordination and speed skills required by operational functions as major impediments to the presence of people with physical disabilities. The absence of this identity group in administrative functions was associated with the notorious access problems in the offices (mainly the lack of ramps and large elevators), which are located in an old building. Nevertheless, the company employs some individuals with light mental disorders in functions that do not involve direct client contact, describing such measure "as part of the company's social responsibility commitment" (Portuguese female HR director).

Blind individuals and wheelchair users collaborate with F4 through telecommuting arrangements because this policy "reduces our expenses with appropriate equipment and space" (Portuguese female HR director). However, since these individuals are kept apart from colleagues and clients, they may have fewer career opportunities [38].

Access ramps were recently installed in F5 to accommodate a worker who became paraplegic after a car accident. Apart from this reactive measure, no formal policies and/or practices concerning people with disabilities were adopted.

F6 employs a number of ear-impaired women in assembly line functions because "they reveal more concentration and are thus less prone to making mistakes" (Portuguese female HR manager). In contrast, people with mobility handicaps are excluded since "the assembly line functions demand high coordination and speed skills" (Portuguese male line manager).

Appearance

Websites from both national and foreign companies usually display pictures of men and women that, according to the traditional beauty patterns, can be considered attractive. In three of the analyzed companies, appearance does seem to matter. Less common looks are not free from judgment at N1, as revealed by a Portuguese male manager: "There is a delivery guy with a funny multicoloured hair and people get suspicious. But, after all, he proved competent".

N2 managers consider less usual appearances (e.g. piercings, tattoos, and peculiar haircuts) as inappropriate for employees dealing with clients because "they do not seem so trustworthy, they don't reflect the company's image" (Portuguese male middle manager). Also in this company, women who want to succeed are expected to renounce their femininity and adopt a more conservative or even masculine look:

> Women who want to succeed here must adjust to a male business world. [...] If you come to a board meeting wearing something revealing, your ideas won't be taken seriously. (Portuguese female top manager)

At F3, less attractive men and women are placed in functions that involve no contact with clients, such as cleaning. In this respect, a top manager (Portuguese male) asserted: "Let's face it: people don't want to be assisted by someone who looks very unattractive or sick!"

Managers from the remaining companies (F4, F5, and F6) did not disclose any particular position regarding physical appearance, seemingly neither valuing nor depreciating diversity at this level.

Religion and Political Ideology

Most of the 24 analyzed websites (i.e. regardless of the companies' nationality of origin) declare that employees are not evaluated on the basis of their religion. Conversely, managers from all of the six further investigated companies considered wiser to keep religion and political ideology private in order "to avoid conflicts, particularly among workers who are less educated and open to difference" (N2, Portuguese female middle manager).

Sexual/Affective Orientation

The websites of companies from foreign origin frequently include sexual orientation among the list of identity attributes that must be protected against discrimination, while this is less frequent on native companies' websites. Individuals from most of the companies participating in stage 2 of the study (N1, N2, F3, F5) vehemently expressed the idea that homosexuality must not be revealed at the workplace:

> That thing [homosexuality] should be kept private. (N1, Portuguese male worker).

> I don't care if a collaborator is homosexual and, especially, I don't want to know. In some cases, you know, ignorance is a bliss. (F3, Portuguese male manager)

However, the same individuals did not consider inappropriate behavior when heterosexual people disclose private matters at work, such as wedding plans, because "these are normal" (N2, Portuguese male middle manager). Moreover, some individuals denied the existence of homosexual collaborators in their company. This result (unlikely to be correct in such companies employing a large workforce) suggests that collaborators who have this sexual/affective identity may be concealing it and, consequently, experiencing anxiety at the workplace [14].

Discussion

Gender emerges as the most frequently mentioned identity attribute on the diversity contents of the 24 corporate websites. Websites from native companies usually declare the organizational commitment to guarantee equal opportunities between men and women, while several websites from foreign companies present gender balance as a relevant goal. Work-life balance initiatives are deemed useful to promote the presence of women at the several hierarchical levels, including management positions, by half of the companies participating in stage 2 of the study. Gender segregation is still a reality in some of the sample companies and this situation is perceived not so much as a problem but rather as something "natural", i.e. associated with tradition and/or physical complexion. Both the persistent gender segregation and the assumption that work-life balance initiatives are mostly developed because women need them reinforce this group's minor status [39].

Websites from both national and foreign companies usually represent collaborators and clients/consumers through pictures of men and women with a complexion that is easily associated to Portuguese native people. Seemingly, these companies devalue the fact that a more diverse population characterizes the contemporary Portugal. Moreover, this option may risk communicating that individuals from national origin are privileged candidates, as well as deemed more significant buyers. On the other hand, pictures of people from ethnic minorities tend to figure on the CSR sections/links, thus suggesting that companies regard them as disadvantaged groups that must be protected for ethic reasons, rather than as a valuable workforce or client pool.

The foreign nationality of origin is faced differently depending on the specific country of origin. Members from more developed countries, such as the US, Germany and the UK, are appraised as possessing more sophisticated know-how and, thus, as significantly contributing to the organizational competitiveness.

In contrast, individuals from Brazil, Angola and Mozambique (which were former Portuguese colonies), as well as from Eastern Europe, are generally found in less intellectually demanding functions. These results suggest ethnical stereotypes and ethnocentric attitudes that echo both the Portuguese colonial past and the world's geopolitical division [40]. As argued by Mighty [9], the individual's geographic origin influences the place s/he occupies at work.

Common age stereotypes in the Western societies [41] seem to prevail and influence people's management decisions in a couple of the analyzed organizations, curiously one native company and an affiliate from foreign origin. Even more usual is managers' perception that the demanding requirements of operational functions are among the major factors preventing the employment of people with disabilities in the correspondent companies. This argument suggests that these individuals are associated with functions that are less intellectually demanding, as if their physical condition implied cognitive problems as well. Moreover, some managers considered wiser to keep individuals from this identity group (as well as less attractive workers) in functions that do not involve contact with external stakeholders. Findings also suggest that these individuals tend to be perceived and dealt with as targets of CSR actions rather than valued by their specific talents. Both practices preclude the full integration of people with psychical disabilities and limit their career opportunities.

Religion and political affiliation are considered controversial subjects that should be silenced at the workplace. Moreover, although it is considered normal and expected that heterosexual employees reveal private life details, homosexuality is perceived as a private issue that should not be disclosed, a view that conceives homosexuality as a deviant behavior.

Drawn on Dass and Parker's [8] typology, Table 3 summarizes the way the companies participating on the study are evaluating and dealing with distinct identity groups.

Table 3 Organizational approaches towards specific identity groups in the analyzed companies

According to Dass and Parker's [8] typology	Main results	Minority identity groups/attributes and corresponding organizations
Resistance Perspective (Diversity as threat)	*Rejection/avoidance* Workplace discrimination. Limited employment opportunities. Organizations do not benefit from minorities' talents and skills	• Ageing individuals (N2, F3) • Unusual/unattractive looks (N2, F3) • Mobility disability and health problems (N1, F6)
	Denial/nonissue Differences must be silenced/concealed. Minorities may experience anxiety and reveal lower performance	• Minority sexual orientations (N1, N2, F3, F5) • Religion and political affiliation (N1, N2, F3, F4, F5, F6)

(continued)

Table 3 (continued)

According to Dass and Parker's [8] typology	Main results	Minority identity groups/attributes and corresponding organizations
Discrimination and Fairness Perspective (Differences cause problems)	*Accommodation/Controlled access* Limited opportunities and access to specific functions. Minorities may be perceived as targets of CSR actions but are not valued by their talent	• Gender segregation (F3, F5) • Physical disability and health problems (N2, F3, F4, F5, F6) • Unusual/unattractive looks (N2, F3) • Immigrants from less developed countries (N1, N2, F3, F5)
Access and Legitimacy Perspective (Differences create opportunities)	*Appreciation* The presence of minorities is valued but not enough to compel organizations to actively promote it	• Gender diversity (N1) • Age diversity (N1) • Immigrants from more developed countries (N2, F3, F5)
	Promotion Minorities are viewed as possessing specific talents/skills due to their identity groups. Their presence is promoted in specific functions	• Gender diversity (N2, F4, F6) • Immigrants from specific nationalities of origin (N2, F4, F6) • Physical disability—the case of ear-impaired women (F6)
Learning Perspective (Differences and similarities offer opportunities and bear costs)	*Full access* Minorities are evaluated on the basis of their particular talents/skills (regardless of identity attributes)	Discourse found among some websites of companies from foreign ownership/control but not among the analyzed companies

Final Considerations

This study particularly contributes by exposing diversity as a social construct that can reflect as many facets as the minority identity groups present in a specific space and time. Most of the diversity approaches described on Dass and Parker's [8] typology were, indeed, found coexisting in the Portuguese organizational context, some of them revealing more than one form. Companies that either reject/avoid the presence of certain minorities (e.g. ageing individuals) or deny the existence of some (less visible) identity attributes among their workforce (e.g. homosexuality) can be considered as adopting the *Resistance Perspective*. Attributes such as physical disability, less common looks, and ethnicity were tolerated but perceived as bringing potential problems to some companies. This evaluation features in the *Discrimination and Fairness Perspective*, which is mainly characterized by confining such minorities to specific jobs and functions, as well as demanding their unilateral adaptation to the organizational culture. The *Access and Legitimacy Perspective*, which implies that differences are perceived as generating specific benefits, was seldom observed in the sampled companies and, when identified, it was mostly circumvented to gender balance and foreign nationality (particularly when referring to more developed countries). Moreover, this approach was found as assuming either a more active form if the organization really develops policies and practices to promote such diversity, or a passive one if the organization merely appreciates the (accidentally) existing diversity. The *Learning Perspective*, which conveys a wider conception of difference and the pursuit of multiple objectives derived from diversity, was detected in the website contents of some affiliates from foreign origin, but was virtually absent from the interviews conducted on the analyzed companies.

This study also exposes diversity as a concept that is socially (re)interpreted and (re)negotiated over time. In concrete, although compassion, solidarity and support are communal values traditionally attributed to the Portuguese society [42], the persistent economic difficulties faced by this country seem to be thwarting the sympathy towards the specific challenges faced by ethnic minorities, since native people evaluate their own situation as equally tough. This is particularly evident when it comes to foreign individuals from less developed countries, such as the former Portuguese colonies in Africa, who are perceived as absorbing job opportunities but failing to contribute with valuable talent. Moreover, most managers considered wise to avoid religious and ideological comments at the workplace, an opinion that may be intensified by the current climate of tensions in Europe. Also significant is the fact that the affiliate from foreign origin with more global activity (F4) distinguishes itself by its appreciation of diversity, particularly in terms of cultural background and gender. Such result demonstrates the impact that complying with distinct institutional contexts may produce in terms of diversity perspectives and practices.

Longitudinal studies that analyze the corporate websites' contents, and particularly the six sample companies, may expose the impact of more current changes, such as the integration of Syrian refugees into Portuguese workplaces and the recently elected left-wing government along with its corresponding social and work policies. On the other hand, similar studies focusing on small and medium-sized enterprises constitute a significant avenue of research since they are still dominant in the Portuguese organizational context. Given that the current study reveals that the same identity group can be perceived differently in distinct organizations operating in the same national context (e.g. avoidance of elder employees vs. appreciation of age diversity), it is expected that such variety of responses would be amplified if companies of different sizes were included. Studies comparing companies from more traditional versus more high-tech activity sectors are also expected to offer stimulating results and discussions, as well as to contribute to a more comprehensive understanding of the way minority identity groups are evaluated and managed in the Portuguese organizational context.

References

1. Barbosa, I., & Cabral-Cardoso, C. (2010). Equality and diversity rhetoric: One size fits all? Globalization and the Portuguese context. *Equality, Diversity and Inclusion: An International Journal, 29*(1), 97–112.
2. Barbosa, B., & Cabral-Cardoso, C. (2014). Words fly quicker than actions: The globalization of the diversity discourse. In G. S. Drori, M. A. Höllerer, & P. Walgenbach (Eds.), *Global themes and local variations in organization and management: Perspectives on glocalization* (pp. 146–160). New York: Routledge.
3. Fernandes, E., & Cabral-Cardoso, C. (2003). Gender asymmetries and the manager stereotype among management students. *Women in Management Review, 18*(1/2), 77–87.
4. Santos, G. G. (2015). Narratives about work and family life among Portuguese academics. *Gender, Work & Organization, 22*(1), 1–15.
5. Santos, G. G., & Cabral-Cardoso, C. (2008). Work-family culture in academia: A gendered view of work-family conflict and coping strategies. *Gender in Management: An International Journal, 23*(6), 442–457.
6. Elmes, M., & Connelley, D. L. (1997). Dreams of diversity and the realities of intergroup relations in organizations. In P. Prasad, A. J. Mills, M. Elmes, & A. Prasad (Eds.), *Managing the organizational melting pot: Dilemmas of workplace diversity* (pp. 148–167). Thousand Oaks: Sage.
7. Milliken, F. J., & Martins, L. L. (1996). Searching for common threads: Understanding the multiple effects of diversity in organizational groups. *The Academy of Management Review, 21*(2), 402–433.
8. Dass, P., & Parker, B. (1999). Strategies for managing human resource diversity: From resistance to learning. *The Academy of Management Executive, 13*(2), 68–80.
9. Mighty, E. J. (1997). Triple jeopardy: Immigrant women of color in the labor force. In P. Prasad, A. J. Mills, M. Elmes, & A. Prasad (Eds.), *Managing the organizational melting pot: Dilemmas of workplace diversity* (pp. 312–339). Thousand Oaks: Sage.
10. Wentling, R. M., & Palma-Rivas, N. (1998). Current status and future trends of diversity initiatives in the workplace: Diversity experts' perspective. *Human Resource Development Quarterly, 9*(3), 235–253.
11. Weldon, S. A. (2006). The institutional context of tolerance for ethnic minorities: A comparative, multilevel analysis of Western Europe. *American Journal of Political Science, 50*(2), 331–349.
12. Acker, J. (2006). Inequality regimes: Gender, class, and race in organizations. *Gender & Society, 20*(4), 441–464.
13. Brickson, S. (2000). The impact of identity orientation on individual and organizational outcomes in demographically diverse settings. *Academy of Management Review, 25*(1), 82–101.
14. Clair, J. A., Beatty, J. E., & MacLean, T. L. (2005). Out of sign but not out of mind: Managing invisible social identities in the workplace. *The Academy of Management Journal, 30*(1), 78–95.
15. Harrison, D. A., Price, K. H., & Bell, M. P. (1998). Beyond relational demography: Time and effects of surface- and deep-level diversity on work group cohesion. *The Academy of Management Journal, 41*(1), 96–107.
16. Litvin, B. R. (1997). The discourse of diversity: From biology to management. *Organization, 4*(2), 187–209.
17. Feldman-Bianco, B. (2001). Colonialism as a continuing project: The Portuguese experience. *Identities, 8*(4), 477–482.
18. Pilkington, A. (2001). Beyond racial dualism: Racial disadvantage and ethnic diversity in the labour market. In M. Noon & E. Ogbonna (Eds.), *Equality, diversity and disadvantage in employment* (pp. 172–189). New York: Palgrave.
19. Herriot, P., & Pemberton, C. (1995). *Competitive advantage through diversity*. London: Sage.

20. Liff, S., & Wajcman, J. (1996). 'Sameness' and 'difference' revisited: Which way forward for equal opportunity initiatives? *Journal of Management Studies, 33*(1), 79–94.
21. Dickens, L. (1999). Beyond the business case: A three-pronged approach to equality action. *Human Resource Management Journal, 9*(1), 9–19.
22. Clarke, L., Pedersen, E. F., Michielsens, E., & Susman, B. (2005). The European construction social partners: Gender equality in theory and practice. *European Journal of Industrial Relations, 11*(2), 151–177.
23. Hodges-Aeberhard, J. (1999). Affirmative action in employment: Recent court approaches to a difficult concept. *International Labour Review, 138*(3), 247–272.
24. Watson, W., Kumar, K., & Michaelsen, L. K. (1993). Cultural diversity's impact on interaction process and performance: Comparing homogeneous and diverse task groups. *Academy of Management Journal, 36*(30), 590–602.
25. Heres, L., & Benschop, Y. (2010). Taming diversity: An exploratory study on the travel of a management fashion. *Equality, Diversity and Inclusion: An International Journal, 29*(5), 436–457.
26. Prasad, A., & Elmes, M. (1997). Issues in the management of workplace diversity. In P. Prasad, A. J. Mills, M. Elmes, & A. Prasad (Eds.), *Managing the organizational melting pot: Dilemmas of workplace diversity* (pp. 367–375). Thousand Oaks: Sage.
27. European Commission. *Justice—Tackling discrimination: diversity management*. http://ec.europa.eu/justice/discrimination/diversity/index_en.htm. Accessed 25 July 2016.
28. Boxenbaum, E. (2006). Lost in translation: The making of Danish diversity management. *American Behavioral Scientist, 49*(7), 939–948.
29. Point, S., & Singh, V. (2003). Defining and dimensionalising diversity: Evidence from corporate websites across Europe. *European Management Journal, 21*(6), 750–761.
30. Risberg, A., & Søderberg, A. (2008). Translating a management concept: Diversity management in Denmark. *Gender in Management: An international Journal, 23*(6), 426–441.
31. Süb, S., & Kleiner, M. (2008). Dissemination of diversity management in Germany: A new intuitionalist approach. *European Management Journal, 26*(1), 35–47.
32. Pina-Cabral, J. (2001). Galvão among the cannibals: The emotional constitution of colonial power. *Identities, 8*(4), 483–515.
33. Cunha, M. P. (2005). Adopting or adapting? The tension between local and international mindsets in Portuguese management. *Journal of World Business, 40*(2), 188–202.
34. Inacio, A. P., & Weir, D. (1993). Management in Portugal. In D. J. Hickson (Ed.), *Management in Western Europe* (pp. 191–203). Berlin: Walter de Gruyter.
35. Kogut, B., & Belinky, M. (2008). Comparing small world statistics over time and across countries: An introduction to the special issue comparative and transnational corporate networks. *European Management Review, 5*(1), 1–10.
36. Dix, S. (2009). Religious plurality within a catholic tradition: A study of the Portuguese capital, Lisbon, and a brief comparison with Mainland Portugal. *Religion, 39*(2), 182–193.
37. Commissioner for Human Rights. (2011). *Discrimination on grounds of sexual orientation and gender identity in Europe*. Strasbourg, France: Council of Europe Publishing.
38. Baruch, Y. (2000). Teleworking: Benefits and pitfalls as perceived by professionals and managers. *New Technology, Work and Employment, 15*(1), 34–49.
39. Smithson, J., & Stokoe, E. H. (2005). Discourses of work-life balance: Negotiating 'genderblind' terms in organizations. *Gender, Work & Organization., 12*(2), 147–168.
40. Frenkel, M. (2005). The politics of translation: How state-level political relations affect the cross-national travel of management ideas. *Organization, 12*(2), 275–301.
41. Chiu, W. C. K., Chan, A. W., Snape, E., & Redman, T. (2001). Age stereotypes and discriminatory attitudes towards older workers: An East-West comparison. *Human Relations, 54*(5), 629–661.
42. Manuel, P. C., & Tollefsen, M. N. (2008). Roman Catholicism, secularization and the recovery of traditional communal values: The 1998 and 2007 Referenda on Abortion in Portugal. *South European Society and Politics, 13*(1), 117–129.

Cultural Complexity in Large Organisations

Nick G. Chandler

Abstract This chapter presents the key concepts and perspectives involving cultural diversity in large organisations. The concept of a large homogenous organisations culture is questioned using the case study of a large Higher Educational Institution and arguments are made for adopting a differentiation perspective. With this new perspective, employee subcultures are viewed in terms of diversity and uniformity. The nature of diversity in the long term is considered through the findings of a longitudinal study of the Higher Educational Institution. The diversity in the organisation is considered in terms of intergenerational diversity through a study the student's values and perception. Finally, the strategic effect of complexity in organisations is considered in terms of the evolutionary nature of strategy.

Diversity in Organisational Culture

Basic Terms and Definitions

The **values** are the general criteria, standards or guiding principles that people have and use to determine which types of behaviour, events, situations and outcomes are desirable or undesirable. An individual's values may not be the same as those of the organisation that management maintain, referred to as **espoused values**. Espoused values are seen as a desired state as put forward by management as part of the image of the organisation rather than the values held by all members of the organisation. When a company indicates their core values on their company profile, these are the espoused values which management hopes that the employees also uphold. These espoused values may include **terminal values**, which are the desired end states or outcomes, such as high quality, strong culture, and can often be found in an organisation's mission and/or vision statement. **Instrumental values** refer to the desired modes of behaviour, such as working hard, keeping to deadlines, etc.) and

N.G. Chandler (✉)
Budapest Business School, Budapest Buzogany Utca 10-12, 1149, Hungary
e-mail: chandler.nicholas@uni-bge.hu

C. Machado and J.P. Davim (eds.), *Managing Organizational Diversity*,
DOI 10.1007/978-3-319-54925-5_3

can be found in company policy with topics such as time-keeping, ethics and anti-discrimination.

The **organisational culture** refers to "a pattern of shared basic assumptions that the group learned as it solved its problems of external adaptation and internal integration that has worked well enough to be considered valid and therefore, to be taught to new members as the correct way to perceive, think, and feel in relation to these problems" [1]. Culture may refer to the 'real culture', which concerns the characteristics of the organisation or the 'constructed culture', which concerns people's perceptions of themselves and others as members of the organisation. An organisational culture may be divided in large complex organisations into groups of subcultures. A **subculture** is a subset of an organisation's members who interact regularly with one another, identify themselves as a distinct group within the organisation, share a set of problems and routinely take action on the basis of collective understandings unique to the group [2]. Within each group of subcultures there may be a **dominant culture**, which means that the subculture contains a majority of the members of the organisation or that the subculture is significantly larger when compared to the other subcultures. The **dominant** culture is often associated with a management subculture since management is often seen as 'in charge' and therefore dominant.

The Importance of Organisational Culture

Every organisation has its own way of doing things. Even within the same industry, companies have their own style of working. The beliefs, ideologies, principles and values of an organisation constitute the organisational culture. It can be seen in the architecture, the uniform (or lack of it), social events and even the way employees choose to celebrate. Consider an organisation with an open place office compared to one with closed cubicles or one where a uniform is compulsory to one where jeans and t-shirts are acceptable and employees lounge on bean bags with their lap taps. The organisational culture is more than just the way things are done or the way things look. The organisational culture affects the organisation in the following ways:

- *The culture decides the way employees interact with management.*

When employees have the same way of doing things as management then it is often referred to as a **healthy culture**. The interaction is positive and supportive to both management and employees, resulting in continuous motivation and loyalty.

- *The culture of an organization gives employees a sense of direction at the workplace.*

One of the keys to strategic management is ensuring that all employees understand the values and direction of the organisation so that strategic decisions can be passed clearly and effectively to lowers levels for implementation. The espoused values of management give a basis for all employees to understand not only what the organisation stands for but also how to accomplish daily tasks. In this way, organisational culture is seen as the **glue** that holds employees together with a set of shared values and common direction.

- *The organisational culture is part of the image of the organization.*

Every organisation has its own way of doing things that make it different from all the others, like an **organisational fingerprint**. It gives an identity to the organisation. In this way the organisational culture enables employees to be aware of and understand what the reason is for the organisation's existence and what values and norms are upheld in the organisation.

- *The organisational culture protects itself, like an immune system.*

Employees that behave against the norms and values of the culture will be seen as not fitting in. In many cases organisations ensure that potential employees have the 'correct' values and behaviour at the interview stage and those that do not are not given positions in the organisation. On the other side of the coin, this also means that if change is needed, there may be **resistance to change** if it involves a change in existing values and behaviours.

- *Organisational culture is something organic.*

The values and behaviours of an organisational culture may change in the face of internal and external changes. Organisational culture change is a slow process. It is said that after a merger, it takes at least 7 years for the two cultures to fully merge [3]. However, organisational culture evolve in the face of environmental changes, be they internal or external.

- *The organisational culture indicates the way things are done.*

The culture dictates the way of working that is preferred in a particular organisation. For example, whether collaboration is preferred over competition, an internal focus is preferred over an external focus or if stability and control is preferred over flexibility and creative freedom.

Given the importance of organisational culture, small business owners need to consider the type of culture they wish to develop in the early stages of establishing a business. Then as the company grows and expands, the top management need to build and reinforce the values and norms held in the organisation so that the culture can become a strong culture. A strong culture is a valuable asset and has additional benefits to the ones already listed in this section:

(1) Attracting the best employees

When an organisational culture is strong, employees are motivated and engaged. This makes the organisation attractive to other workers in the industry who perhaps do not have such a strong culture. In industries where talent is in short supply and many organisations offer similar compensation packages, the organisational culture may be the deciding factor for talented workers to apply for a job. After all, a strong culture means that employees enjoy working at the organisation and believe in the direction in which the company is heading.

(2) Holding onto the best employees

If employees are motivated, engaged and believe in the core values and the desired direction or **vision** of the organisation, then they are less likely to look for a job somewhere else. As mentioned earlier, organisational culture is the glue that hold employees together the stronger the culture, the stronger the glue holding them together.

(3) No more daily chore

When a strong culture exists and people are engaged and motivated, then the job is no longer seen as a daily list of chores that have to be done so that a salary is paid. People no longer toil to get through the day but comes to work because they enjoy it. Organisations with strong cultures are likely to have lower absentee rates through sickness and other causes, as well as a lower number of staff leaving the organisation, i.e. a lower **staff turnover**.

Building a strong culture requires some investment in time and money, but the results significantly outweigh the costs. The culture needs to be considered from day one and continuously built upon. However, as the organisation becomes large and increasingly complex, top management need to consider the likelihood that the organisational culture is no longer a single homogenous culture but may be fragmented into smaller cultural groups or subcultures. The following section considers the organisational culture of large organisations.

Complexity in Organisations

When we look at society we see a wide range of values, ages, behaviours and attitudes. If we consider large complex organisations then we can assume that such organisations reflect a similar complexity as the society in which they operate [4]. This means that in large organisations there are likely to be a diverse range of values, behaviours, attitudes and perceptions.

If we see large organisations as complex organisms, then this also means the only way to fully understand them is to find the differences across the organisation and analyse each diverse part, in other words: “it is only by understanding the parts …we can understand the whole” [5]. Accepting diversity and complexity in this context seems relatively straightforward. However, this acceptance only works if we consider all the members of the organisation—not only the management but all levels of the organisational structure. The greater size of an organisation means the greater the likelihood of it splitting into subcultures. An example of the diversity of subcultures can be seen in the following table:

Table 1 Subcultures' characteristics in a large organisation

	Return culture	Market culture	Profession culture	Small labourers
Members	Product managers [top managers (to some extent); potentially: finance]	Sales (potentially: customer care)	Technicians (to some extent: the lawyer)	Invoicing, MIRA, lawyer, customer care, finance
Self portrait	The conducting midfielders	The magic forwards delivering goals	Libero, defender serving the others	Secret talents on the bench
Perception of others	Skillful gamblers	Over occupied little star alike	Overloaded geniuses somewhere in the building	Ambitious ballasts
Internal—external focus	Intermediate internal	Strong external (customers)	Intermediate external (suppliers)	Miscellaneous (potentially internal)
Attitude towards risk	Intermediate	Risk taker	Risk avoider	Risk avoider
Time orientation	Intermediate	Shorter	Longer	Intermediate–longer
Professional—task orientation	Task orientation	Task orientation	Professional orientation	Task orientation (some professional)
Professional—business orientation	Business	More business than professional	Professional	Professional

Source Bokor [6]

Table 1 shows that an organisation does not have to be an international organisation to experience diversity. Through subcultures, employees see themselves and others very differently. They have varying attitudes to risk, time and whether they focus on internal operations or the external environment. However, it cannot be assumed that a large organisation means a large number of subcultures. The view taken of an organisation's culture is referred to as the cultural perspective and the different perspectives will be shown in the following section.

Cultural Perspectives

Not all experts agree on the extent of diversity in organisations. Perspectives on organisational culture vary from the one end of the scale where all employees have the same or similar values and beliefs, to the other end of the scale where none of the employees share the same belief and values. The way we see or chose to see organisational culture in terms of the level of diversity is referred to as the **cultural perspective**. There are three perspectives:

The first is called the unitarist or **integration** perspective. In this view, we believe that there is unity throughout the organisation. This means that we can classify each organisational culture as a particular type, as with Handy [7] and the four culture types: task, power, people and role-oriented cultures, or Hofstede [8] with an organisation having a role, achievement, power or support culture. There are a number of assumptions that we also have to accept if we adopt this perspective. First, leadership is top-down, with members unified and conforming to the directives of top management. There is a high degree of homogeneity in terms of values, attitudes and behaviours, which is why this perspective is also often referred to as the 'integration perspective' [9]. The shared values and beliefs are seen by members as 'the way we do things round here' [10], rather than 'the way some of us…' or 'the way most of us do things around here'. It seems reasonable for managers to assume the integration/unitarist perspective as this reinforces their desire for all staff to 'tow the line' and ties in with the concepts of vision and mission as an integrative force encouraging improved staff performance and increased unity of direction.

The second option is referred to as the pluralist or **differentiation** perspective. In this case, diversity is accepted to some extent and is reflected in the existence of diverse subcultures in organisations. The culture is no longer homogenous, but rather 'heterogeneous'. Even in smaller organisations such as supermarkets, subcultures have been found to exist [11]. Conformity towards a single monolithic organisational culture is replaced by cultural diversity and the potential conflict between these subcultures is tolerated as management takes a 'multicultural view' of their organisation.

The final perspective sees the organisation and its members in an even greater state of diversity. The organisation consists of individuals with their own values and norms and as such neither a single dominant culture nor any subcultures are said to exist. In such organisations managing cultural change is impossible on an individual basis and the focus shifts towards communication and diversity management. This perspective is referred to as the **fragmentation** perspective with fragmented groups being issue-specific and no shared meaning between members of the organisation or members of part of the organisation. The organisation appears as a series of contradictions and confusion on the part of the members with a lack of consistency, consensus and ambiguity.[1]

Whichever perspective is adopted, the caveat remains that this perspective only serves when taking a snapshot of the organisation at one particular moment in time as values are neither completely stable nor unstable but rather change according to the environment of individuals and groups [12].

[1]Martin [9], ibid.

Levels of Cultural Diversity

Subcultures are often understood to be countercultures in organisations and often seen in a negative light. However, subcultures do not automatically translate to 'countercultures', rebelling against the organisation's top management. Subcultures may exist alongside a dominant culture [13] as each subculture has members with a combination of pivotal and peripheral values. **Pivotal** values are the core values of an organisation as espoused by top management. Members are expected to uphold these values, with those that do not being rejected, or ejected, from the organisation [14]. **Peripheral** values refer to those values not considered as core values but the organisation may encourage members to adopt them. Failure to take on the peripheral values, however, does not result in rejection in the same way as when failing to uphold the pivotal values. When members of a subculture have the same pivotal and peripheral values as the overarching organisational culture (i.e. the dominant culture), then it is called an **enhancing** subculture. When the subculture and dominant culture shares the pivotal but both the peripheral values, then it is referred to as an **orthogonal** culture. Finally, when the subculture does not share the pivotal or the peripheral values of the dominant culture, then it is called a **counterculture** [15]. Even this breakdown of complex cultures into subcultures with a combination of pivotal and peripheral values is further complicated by the view that subculture types may be seen on a scale of increasing diversification rather than as three concrete types, as can be seen in Fig. 1 [16].

As can be seen in the figure, a unitary culture refers to the integration or unitarist perspective with a single monolithic organisational culture. An integrated organisational culture is when the enhancing subcultures are a part of the overall dominant organisational culture which may be seen as a combination of the integration and differentiation perspectives. The slightly differentiated organisational culture refers to a collection of both enhancing and orthogonal subcultures, with varying combination of peripheral and pivotal values and still takes a combination of the integration and differentiation perspectives. A significantly differentiated culture refers to no enhancing subcultures and only orthogonal or counter subcultures. The subcultures may be heterogeneous (a differentiation perspective), but there is still the existence of a dominant culture as well. In the disorganised form of organisation, there is no dominant culture and subcultures have no common values, which takes the fragmentation perspective. It now seems that a fourth perspective has emerged: the **multi-perspective approach** where the cultural composition of an organisation could be one of a number of possible combinations with varying degrees of common peripheral and pivotal values for enhancing and orthogonal subcultures, countercultures, a dominant culture and fragmented sections of ambiguity and uncertainty.

It has been argued that in some organisations, such as prisons, pivotal values are so widely adopted that they restrict the emergence of peripheral values and thereby, the emergence of subcultures.[2] However, this is unlikely in strong culture

[2]Schein [13], ibid.

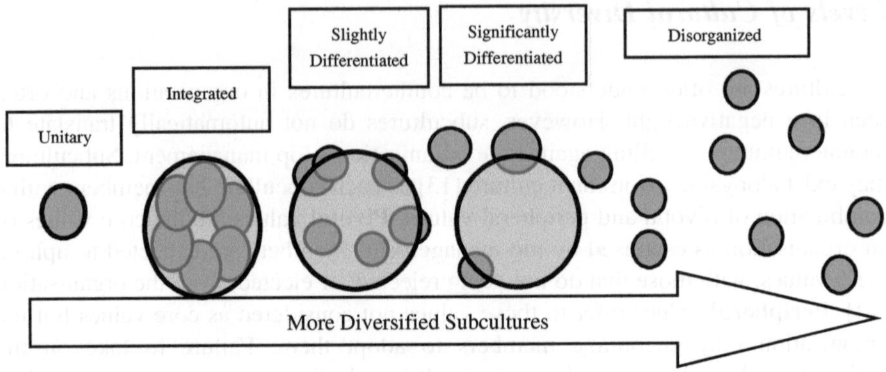

Fig. 1 The diversification of subcultures. *Source* Hatch [16]

organisations. The extent that employees have pivotal or peripheral values in line with the organisation not only indicates if the employee (or subculture) are enhancing the values and beliefs of the organisation, but also the way the employees (or subcultures) view the organisation and the world around them. This is referred to as cultural lenses. In a large organisation made up of people from different nationalities, employees learn to see things through the eyes of others as they learn and appreciate cultural differences. However, in organisational culture can still be diverse without national differences and the way employees perceive the organisation, its management and the world around them can impact upon their individual motivation and effectiveness.

Let us take an example of subcultures in a large organisation. A national company with little diversity on an international level. However, employees have grouped to form subcultures identified as certain types, such as a market (externally focussed with stability and control) subculture, a hierarchy (formal procedures, bureaucracy, internal focus with stability and control) and clan (family and group centred with an internal focus as well as flexibility and a sense of freedom). These subcultures exist with some harmony in the organisation and have common pivotal values, as the clan and hierarchy subcultures have a common internal focus and integration and the dominant market culture and hierarchy culture have common values of stability and control. Thus management need to understand that, like links in a chain, the hierarchy subculture has pivotal values in line with the dominant culture, and that although the clan subcultures had no pivotal values in line with the market culture, this subculture was not entirely isolated in the organisation as it has pivotal values in line with the hierarchy subculture, as seen in Fig. 2.

It can also be seen from the above figure, there is a combination of perspectives in play. We have integration (a dominant culture), differentiation (competing subcultures) and fragmentation (residuals that do not fit any category). Thus diversity is in fact represented in large organisation with a combination of these perspectives or levels rather than a choice of one out of the three. There are examples of enhancing (hierarchy subcultures) and orthogonal (clan and market subcultures) in relation to the organisation's hierarchy culture.

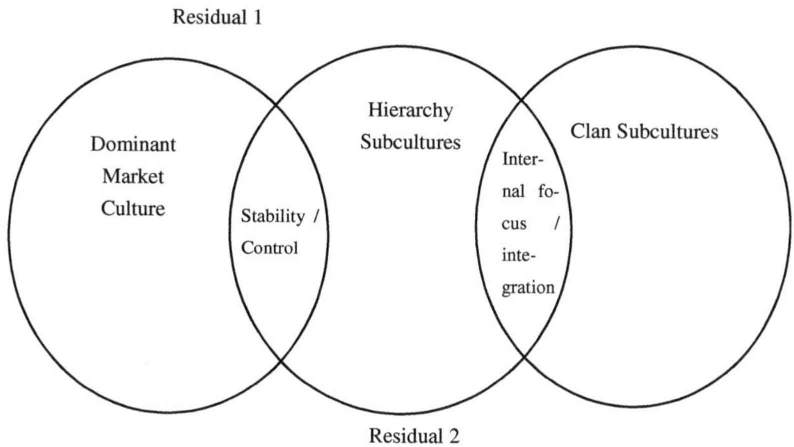

Fig. 2 The composition of culture in the organisation

The Implications

To consider the implications of the potential cultural and intergenerational diversity in large organisations, a specific case may serve as an example of the distinct complexity: the complexity of culture in higher education has been covered for almost half a century. The organisational culture in higher educational institutions (HEIs) has been described as 'tribes and territories'.[3] The potential for diversity is high, with teaching staff experiencing high autonomy, indicating a lower likelihood of shared values and yet a high respect for traditions and rituals which may indicate a strong culture. The question arises: is the culture in higher education split by profession, with teaching and non-teaching staff, are teaching staff split by topics taught? Is there a difference in culture for teachers of soft or hard subjects? The following section will consider the case of diversity in organisations in higher education.

Cultural Diversity in Higher Education

A number of studies of organisational culture in higher education have uncovered the potential for accepting two of the three perspectives mentioned earlier in this chapter: integration and differentiation [17] and this is referred to as the multi-perspective approach. The unitarist perspective seems plausible when we think of faculty members all sharing a common world view and the same approaches to learning and the academic world in general. Likewise, attitudes to

[3]Becher [5], ibid.

research and development may also indicate the existence of shared values and a common desired direction for the organisation. However, the majority of studies indicate that cultures of HEIs are complex and rarely homogenous [18] and that in these institutions subcultures will emerge. In the example mentioned earlier, if faculty share similar values towards research, learning and the academic world in general, then rather than being an argument in favour of homogeneity in cultures, it is in fact an indicator of the existence of **occupational subcultures**. Higher education organisational cultures are differentiated from others because of greater complexity [19]. Beyond occupational subcultures, departments are often split according to the discipline. These departments may also be separated according to location and may feel the need to compete with other departments for funding for the yearly budgets, numbers of students enrolled on courses, number of publications or quality of research by staff. These competing aspects all serve to form **cultural boundaries**. The resulting effect is that subcultures form based on the discipline or specialisation of staff. Universities are often made up of a collection of institutes or Faculties that may exist in different locations and may be seen as separate units developing their own cultures. Furthermore, in higher education there is also a student culture which may challenge and even change the values of the organisational culture, and vice versa. This cultural complexity can be seen in Fig. 3, which is referred to as a **cultural web**.

Subcultures are also more likely to develop in bureaucratic, larger or more complex organisations with a wide range of functions and technologies [20]. There are a variety of types of organisational subcultures, not all of which are based on expressing opposing views,[4] as in the case of orthogonal and enhancing organisational subcultures where some values and norms may differ from those of the dominant culture, but there is still adherence to the core or pivotal values. Thus, these subcultures do not impede organisational performance through conflict and resistance to organisational values and norms. In an organisation with heterogeneous subcultures, competing subcultures may cause conflict but the competition between the subcultures may enhance members' roles in the organisation, as, for example, they strive to acquire more skills than the members of other subcultures [21]. In a higher education setting, this competitive aspect could be seen in the number of papers produced per department or rivalries based upon prestige or reputation.

When considering the potential impact of occupational subcultures, conflicts are often seen as the likely outcome. Conflicts have arisen between managerial subcultures which aim to control work within the organisation and other occupational subcultures that seek autonomy [23]. Subcultures are often characterised as 'containing seeds of conflict' as conflict may emerge when members of differing subcultures confront one another [24] due to differences in values, attitudes and behaviours. However, subcultures may also perceive themselves and other subcultures and members in the organisation differently. By subcultures seeing events

[4]Trice and Beyer [20], ibid.

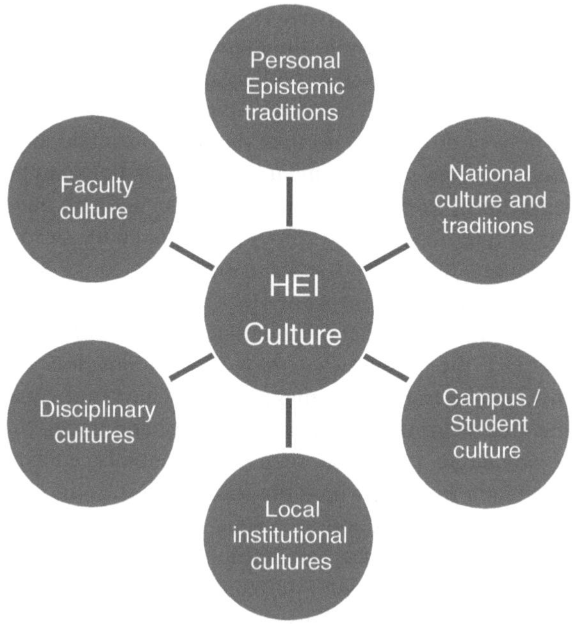

Fig. 3 Cultural complexity in higher education. *Source* Adapted from Valimaa [22] and Becher [5]

through their own 'subcultural lenses' may also be a cause of conflict[5] noted that in multicultural organisations; members of subcultures perceived things only from their cultural perspective (ethnocentrism), also perpetuating conflict.

The cultural diversity found in organisations in the form of subcultures does not instantly mean that the organisation's member are set for conflict and poorer performance. In fact subcultures may have a positive effect upon strong culture organisations in relation to an organisation's desire to become 'agile' [25]. **Agility** is achieved through subcultures providing the flexibility and responsiveness that a unitary culture may limit. A prerequisite of this is that top management adopt a multiculturalist view in that they allow the emergence and co-existence of heterogeneous subcultures in the organisation. There are a number of other reasons why subcultures may not have a negative impact upon the organisation. First, subcultures may consist of smaller groups that are strategically weak and, therefore, not threatening [26]. Second, subcultures often emerge in response to changing demands and can serve as an outlet for members to express conflict and dissent arising during turbulent times. These emerging subcultures are seen as a mechanism for changing less central values as well as a means by which members can express themselves. Furthermore, subcultures may have a particular focus that makes them specialist in certain business activities [27]. If organisations turn their attention towards examining and understanding subcultures, then this act and the attention given to the organisation will aid the organisation in understanding the heretofore

[5]Gregory [4], ibid.

hidden complexities, reasons for conflict and more generally heighten the management's sensitivity to the cultural implications of business operations in general and change processes in particular. Finally, subcultures may strengthen rather than harm organisational cultures: a contrary point of view can strengthen beliefs, values or behaviour as they are then put to the test and any challenge to existing beliefs is thus a tool for reinforcement of values, beliefs and behaviour. As mentioned earlier, subcultures contain a combination of both pivotal and peripheral values and so subcultures could be orthogonal, enhancing or counter cultures [1]. This is an important distinction: some subcultures could be destructive to the organisation (countercultures), but some may not.

In spite of this list of potential benefits from subcultures, this does not discount the fact the subcultures may in turn develop into countercultures that hinder business operations, damage performance and cause conflict. In today's environment of continuous change, instability and uncertainty, a destabilising element like a subculture could be seen as potentially causing failure of the organisation, especially in cases where the leadership is going through a period of transition and the strategic direction is vague or unspecific, as in the case of organisations undergoing a merger, for example.

In higher education, subcultures can act as "containers of creativity in which ideas can formulate relatively independently of the constraints or influences of the (strong) culture" [28]. The phenomenon of subcultures working alongside a dominant culture has been observed in multiple cultures within organisations and indicated a potential positive effect of these subcultures due to the variations between them, hence coining the term 'ambidextrous organisations' [29]. Earlier it was said that common pivotal values are the key to subcultures existing harmoniously with the dominant culture; however, it would be incorrect to assume that large multi-faculty universities (referred to as a 'multiversity') or even small institutions have something in common or some shared characteristic [30]. In the 1981 funding crisis in the UK this conflict of pivotal values in the professional culture of HEIs was seen when power of veto was held by the faculties and departments and in some cases departments and faculties vetoed against the interests and concerns of their own institutions [31]. In a college or university, this antagonism between subgroups may result in member conflict and so they stop talking resulting in the formation of two distinct subcultures [32], and thereby increasing the complexity and the potential for more conflict further down the line.

Employees may try to modify those values of other employees which seem inappropriate to achieving their own goals or the success of the organisation and this line of thought could be extended to an employee disagreeing with the organisations values whilst continuing to work. This is a severe hindrance to the efficiency of the employee and at the very least would result in the development of **cognitive dissonance**, which a person holding two opposing views at the same time. Such views not only stress the dynamic nature of subcultures but also that there is a tendency to reduce the cultural distance between the subculture and the dominant culture [33]. This is another indication that the impact may not necessarily be a negative one.

Finally, a subculture in any organisation may judge the others behaviour as 'abnormal' as they fail to live up to subcultural norms. The resulting effect is alienation, a higher potential for misunderstandings, lack of appreciation of others and conflict.

The Way Forward

Managing Subcultures

For large complex organisations, unitary cultures are actually the exception to the rule and multiple subcultures should be seen as the norm in organisations and not solely on a management level [34]. When organisations wish to develop a strong culture in large complex organisations with a high likelihood of subcultures, then subcultures may be aligned as a means of strengthening the culture. The following model shows a process by which organisations may seek to strengthen organisational culture through the alignment of subcultures, referred to as a **cultural audit** (Fig. 4).

In the figure, it can be seen that this is a continuous process as it is assumed that cultures and subcultures are dynamic in the organisation and that through

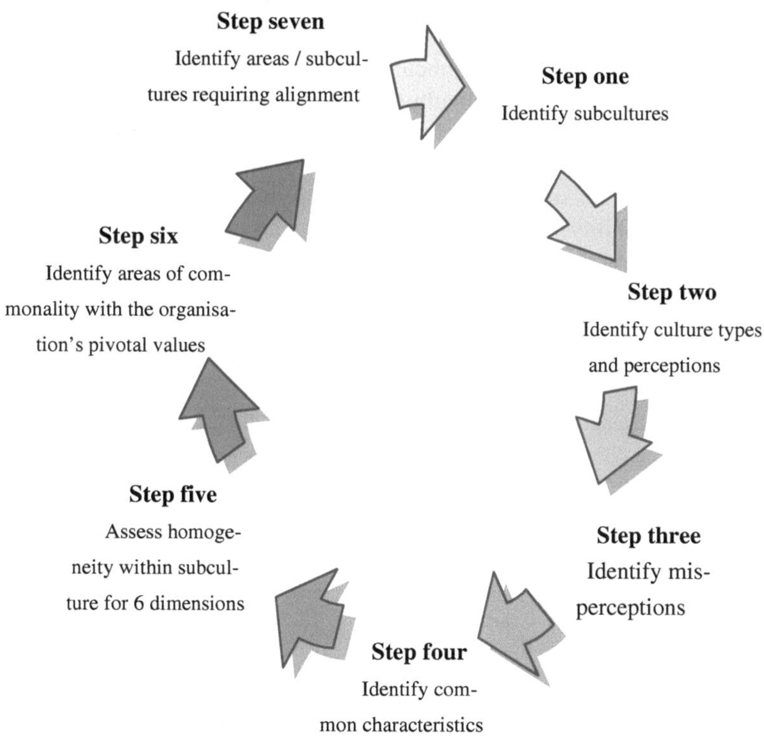

Fig. 4 The change management process for aligning organisational subcultures

interaction, as was suggested in the discussion part of this study, when one subculture changes another may respond in kind either following the new set of values, taking them on partially (as in this case when subculture three expected the leadership to take on a market-culture style of leadership despite being a hierarchy subculture type), or rejecting them partially or fully. Alternatively, this model could be applied in practice as a means of conducting a 'subculture' audit prior to the commencement of any change processes or when looking to implement a change in the direction of the organisation.

Change Over Time

Conducting a subculture audit in a large complex organisation seems a means by which the organisation can achieve a common direction shared through the whole organisation. This is the desired result from a strategic point of view. However, diversity has its benefits and the outcome of the cultural audit does not necessarily result in the alignment of all subcultures. A recent study of the Budapest Business School in 2011 [35] found that each subculture in an organisation may in fact have its own **specialisation**. The organisation was a higher education institution and each subculture had a different focus in terms of its market orientation. As can be seen in the following figure, subcultures with a hierarchy culture type had an internal focus and preference for stability. These subcultures were firmly oriented towards helping and supporting students. The subculture with a market culture type also had a preference for stability and control, but this was combined with an external focus. The market subcultures were oriented towards competition. The third type of subculture was the clan subculture and this group had an internal focus but also a preference for flexibility and freedom of action. The clan subcultures had a preference for integration and a strong cooperation orientation (Fig. 5).

Market orientation in higher education is made up of these three elements (cooperation, competition and student), and so the alignment of these subcultures means that all three subcultures may be pushed towards one particular culture type

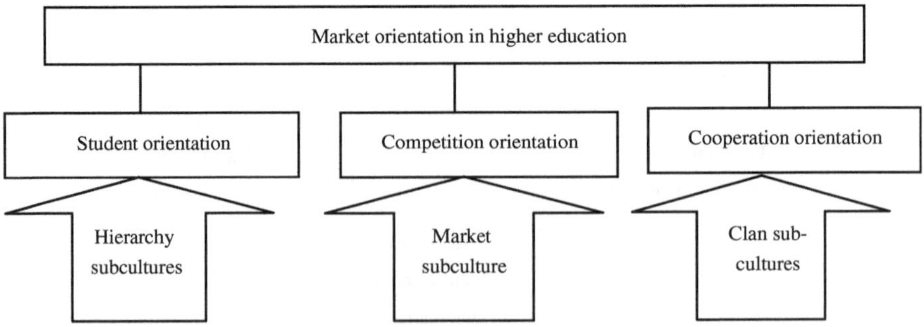

Fig. 5 The contributions of subcultures to market orientation

and one particular orientation. Thus, having a single culture type in the organisation, may lead to a lower overall market orientation and reduce the competitiveness of the organisation. Cultural audits are a means of understanding the values and perceptions of different subcultures in an organisation, but the need for mono- or multiculturalism depends very much on top management conducting a scenario analysis along the lines of asking: 'what if this subculture no longer existed?'

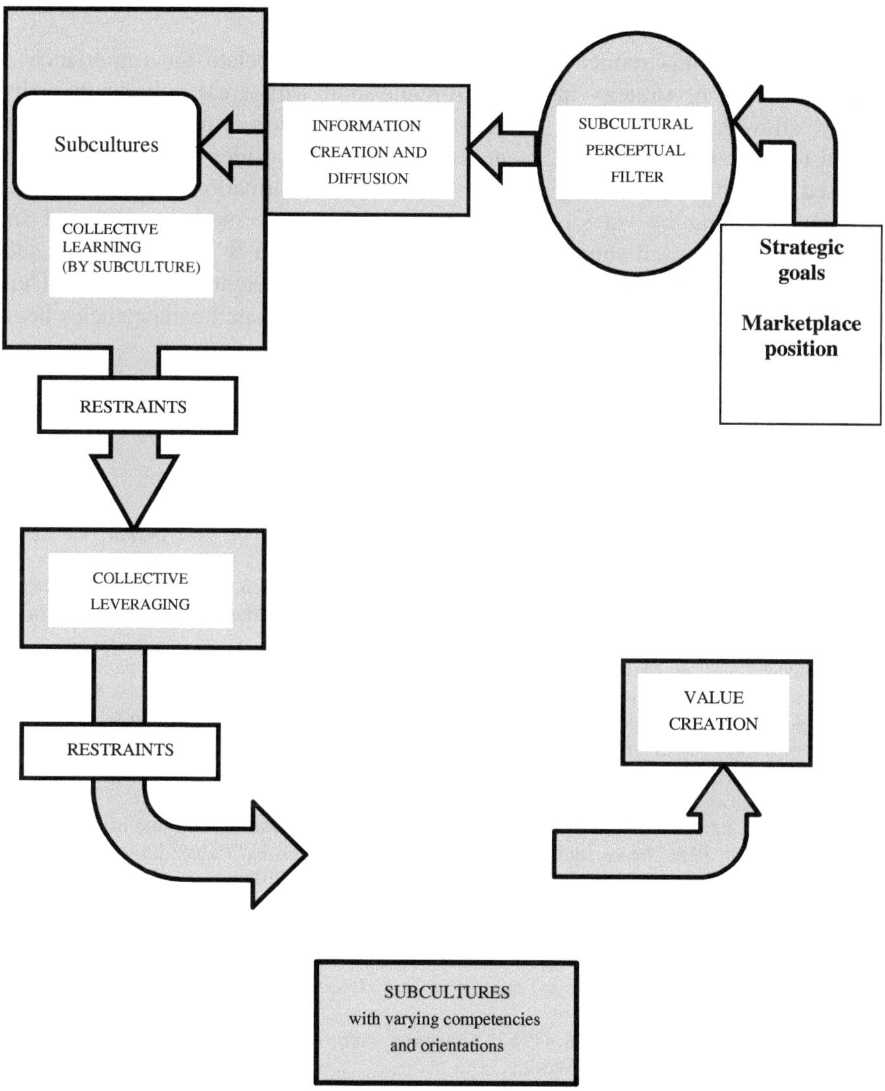

Fig. 6 How strategy unfolds in cultural diversity

Cultural Diversity and the Evolution of Strategy

The uppers tiers of management develop strategic goals based upon key considerations such as marketplace position and capabilities, whilst organisational subcultures evolve organisational competences relating to the information passed on to them concerning the organisation's orientation. However, the information received by top management on strategy is interpreted according to the subculture's view of themselves, others and the greater organisation, called the **cultural perceptual filter** [36] and is similar to the cultural lenses referred to earlier in this chapter. This process can be seen in Fig. 6.

It is through this model that management can appreciate the importance of implementation of strategy in a large organisation with great cultural diversity. Firstly, allowance needs to be made for how information is perceived and interpreted by subcultures. Second, management need to consider how information is diffused, i.e. although strategic plans and related information may be diffused through the hierarchy via top-down communication, information is diffused and interpreted within each subculture. Finally, value creation is specific to each subculture as **collective learning** will produce a range of different competences. Thus, value creation is seen in a range of orientations and associated competencies being covered by each subculture.

References

1. Schein, E. H. (1985). *Organizational Culture and Leadership. A Dynamic View*. San Francisco: Jossey-Bass.
2. Van Maanen, J., & Barley, S. (1985). Cultural organization: Fragments of a theory. In P. J. Frost, L. F. Moore, M. R. Louis, C. C. Lundberg & J. Martin (Eds.), *Organisational culture* (pp. 31–53). Beverly Hills, Sage.
3. Nahavandi, A., & Malekzadeh, M. (1993). *Organizational culture in the management of mergers*. Westport: Quorum Books.
4. Gregory, K. L. (1983). Native-view paradigms: Multiple cultures and culture conflicts in organizations. *Administrative Science Quarterly*, (28), 359–376.
5. Becher, T. (1987). The disciplinary shaping of the profession. In B. R. Clark (Ed.), *The academic profession*. Berkeley, University of California Press.
6. Bokor, A. (2000). Organizational culture and knowledge integration: The case of new product development. *PhD Thesis. Budapest University of Economics and Public Sciences.* Available at: http://phd.okm.gov.hu/disszertaciok/ (Accessed June 15, 2016).
7. Handy, C. (1993). *Understanding organizations* (4th ed.). London-UK: Penguin Books Ltd.
8. Hofstede, G. (1980). *Culture's consequences: International differences in Work-related values*. Beverly Hills, CA: Sage.
9. Martin, J. (1992). *Cultures in organizations: Three perspectives*. New York: Oxford University Press.
10. Deal, T. D., & Kennedy, A. A. (1982). *Corporate cultures: the rites and rituals of corporate life*. Reading, MA: Addison-Wesley.
11. Ogbonna, E., & Wilkinson, B. (1990). Corporate strategy and corporate culture: The view from the checkout. *Personnel Review, 19*(4), 9–15.

12. Seevers, B. S. (2000). Identifying and clarifying organizational values. *Journal of Agriculture Education* [On-line], *41*(3). Available at: http://pubs.aged.tamu.edu (Accessed May 04, 2016).
13. Schein, E. H. (1988). Organizational socialization and the profession of management. *Sloan Management Review*, (30), 53–65.
14. O'Reilly, C., & Chatman, J. (1996). Culture as social control: Corporations, cults and commitment. *Research in Organisational Behaviour*, (18), 157–200.
15. Schein, E. H. (1999). *The corporate culture survival guide*. San Franscisco: Jossey-Bass.
16. Hatch, M. J. (1997). *Organizational culture, organization theory*. New York: Oxford University Press.
17. Metzger, W. (1987). The academic profession in the United States. In B. Clark (Ed.), *The academic profession: National, disciplinary, and institutional settings* (pp. 123–208). Berkeley: University of California Press.
18. Fralinger, B., Olson, V., Pinto-Zipp, G., & DiCorcia, M. (2010). Organizational culture at the university level: A follow-up study using the OCAI instrument. *2010 EABR & ETLC Conference Proceedings*, Dublin, Ireland.
19. Dill, D. D. (1982). The management of academic culture: Notes on the management of meaning and social integration. *Higher Education*, (11), 303–320.
20. Trice, H. M., & Beyer, J. M. (1993). *The culture of work organizations*. Englewood Cliffs, N. J: Prentice-Hall.
21. Sackmann, S. (1992). Culture and subcultures: An analysis of organizational knowledge. *Administrative Science Quarterly, 1,* 140–161.
22. Valimaa, J. (1998). Culture and identity in higher Education research. *Higher Education, 36*, 119–138.
23. Trice, H. M. (1993). *Occupational subcultures in the workplace*. Ithaca, NY: ILR Press.
24. Van Maanen, J., & Barley, S. (1985). Cultural organization: fragments of a theory. In P. J. Frost, L. F. Moore, M. R. Louis, C. C. Lundberg & J. Martin. *Organisational culture* (pp. 31–53). Beverly Hills, Sage.
25. Boisnier, A., & Chatman, J. A. (2002). The role of subcultures in agile organizations. In R. S. Peterson & E. Mannix (Eds.), *Leading and managing people in dynamic organizations* (Chapter 6). Mahwah, NJ: Lawrence Earlbaum Associates.
26. Galunic, D. C., & Eisenhardt, K. M. (2001). Architectural innovation and modular corporate forms. *Academy of Management Journal*, (6), 1229–1249.
27. Heidrich, B., & Chandler, N. (October, 2011). The effect of market-oriented subcultures on post-merger higher education institutions. *US-China Education Review*, *1*(5).
28. Martin, J., & Siehl, C., (1983). Organizational culture and counterculture: An uneasy symbiosis. *Organizational Dynamics*, (122), 52–65.
29. Tushman, M. L., & O'Reilly, C. (1996). Ambidextrous organizations: Managing evolutionary and revolutionary change. *California Management Review, 38*(4), 8–29.
30. Barnett, R. (2000). *Realizing the university in an age of supercomplexity*. Buckingham: Open University Press.
31. Silver, H. (2003). Does university have a culture? *Studies in Higher Education*, *28*(2).
32. Kuh, G., & Whitt, E. (1988). *The invisible tapestry*. Culture in American Colleges and Universities. ASHE-ERIC Higher Education, Report No. 1.
33. Aronson, E. (1968). Dissonance theory: Progress and problems. In R. P. Abelson, E. Aronson, W. J. McGuire, T. M. Newcomb, M. J. Rosenberg, & P. H. Tannenbaum (Eds.), *Theories of cognitive consistency: A sourcebook* (pp. 5–27). Chicago: Rand McNally.
34. Palthe, J., & Kossek, E. (2002). Subcultures and employment modes: translating HR strategy into practice. *Journal of Organizational Change Management, 16*(3), 287–308.
35. Heidrich, B., & Chandler, N. (October, 2011). The effect of market-oriented subcultures on post-merger higher education institutions. *US-China Education Review*, 1(5).
36. Deneault, D., & Gatignon, H. (2000). *An evolutionary theory of firm orientation: from its behavioural to its strategic manifestations*, Fontainebleau, France: INSEAD. Available at: https://flora.insead.edu/fichiersti_wp/inseadwp2000/2000-09.pdf (Accessed July 13, 2016).

Employee and Human Resource Managers Perceptions About Family-Friendly Work Practices: A Case Study Focused on Perceived Organizational Support

Sandra Amorim and Gina Gaio Santos

Abstract This chapter looks at employee's perceptions regarding the type of family-friendly work practices available to them by their employing organizations and, at the same time, it conveys their views on the organizational support received on these matters. The empirical research is based on a case study design that includes four companies belonging to different industries in Portugal. Overall, we conducted 24 in-depth interviews with both operational employees and human resource managers. The findings show that the HR managers in all the companies admit not having formal procedures on family-friendly work practices but only a set of informal practices that vary according to the functional level and the employee rank. The company size and resources, the type of operational activities or the work schedules are important explanatory factors for the scant adoption and implementation of family-friendly practices in these companies, and seem to play a more influential role than institutional forces. Immediate supervisor and co-worker support are perceived by the employees as fundamental sources of organizational support in the work–family life conciliation. The study limitations and future research suggestions are also presented.

Introduction

In a world economy increasingly competitive in terms of work rhythms and work pace and progressively more diverse regarding family-types, such as dual-career couples or single-parent families, the relationship between professional and family

Sandra Amorim holds a Master degree in Human Resources Management from the School of Economics and Management, University of Minho. Presently, she works for a private sector company.

S. Amorim · G.G. Santos (✉)
Department of Management, School of Economics and Management,
University of Minho, Campus Gualtar, 4710-057 Braga, Portugal
e-mail: gaiogina@eeg.uminho.pt

C. Machado and J.P. Davim (eds.), *Managing Organizational Diversity*,
DOI 10.1007/978-3-319-54925-5_4

life has become a central concern for individuals, but also for organizations and society overall. Recognizing that the issue of work and family life encompasses distinct matters related to organizational culture, family trends, and diversity issues, it is important to understand how current organizations manage this issue. Thus, we aim answering the following research questions (RQ):

RQ1 What type of family-friendly work practices and policies can be currently found in employing organizations?

RQ2 What are employees' perceptions about the organizational support regarding the various family-friendly practices (FFPs) that are made available to them by their employers?

The current study intends to shed some light into these matters and it is organized as follows. In the first part, we establish the meaning of work and family life and we conduct a brief literature review on the issue of FFPs and organizational culture. In the second part, we proceed with a description of the methodology used in the empirical study—a case study undertaken with four organizations in the Portuguese context. The third part is focused on the findings presentation. Lastly, we look at some limitations of the research and present suggestions for future research.

Defining the Work–Family Life Relationship

The last decades have been characterized by various economic and social changes that made it difficult to manage the relationship between work and life domains, leading to a growing interest in this subject. It is important to conceptualize what we mean by "work" and "life." Thus, we use Pocock et al. [44: 23] conceptualization that defines work as "paid work" and "life beyond work as the activities outside of paid work including household activities and those activities with family, friends and community, including care activities and voluntary activity." In this research, we are particularly focused in family activities and in the work–life policies that are being implemented by employing organizations in that matter.

The relationship between work and life domains has been conceptualized in various ways, being one of the most widespread the conflict perspective [for a comprehensive review see the studies by 17, 18]. Greenhaus and Beutell [27: 77] define work–family conflict as a "form of inter-role conflict in which the role pressures from the work and family domains are mutually incompatible in some respect." According to the authors, there are three distinct sources of conflict: (1) time-based conflict, where time spent in one role (work) cannot be devoted to activities in the other role (family); (2) strain-based conflict, when strain in one role affects one's performance in the other role and; (3) finally, behavior based-conflict, where patterns of behavior in one role may be incompatible with expectations regarding behaviour in another role. Studies on conflict and interference (sometimes also referred in the literature as negative spillover) are rooted in role theory, which argues that role conflict (and role overload) occurs when there is a simultaneous occurrence of two or more sets of pressures such that the compliance with one would make more difficult the compliance with the other [18].

More recently, Ashforth and colleagues [5] developed the concept of 'boundary theory' to depict the micro-role transitions between distinct life domains (e.g., work and family), involving the crossing of a set of boundaries on a daily basis and along a continuum of segmentation and integration. The boundary theory presupposes that it exists some degree of flexibility and permeability between life domains. As Pocock et al. [44: 24] explain work and life are interactive domains that frequently overlap each other and that are characterized by the spillover between them. Thus, the boundaries between work and family life are not closed but porous and varying in their degree of boundary permeability and porosity [5, 44]. Hence, the porosity between domains may be spatial (e.g., with work being done away from the workplace); temporal (e.g., with work and caring activities being conducted simultaneously); and interactive (e.g., with the actions of one domain affecting the other) [44: 24].

Also important, is the concept of "flexibility enactment" introduced by Kossek et al. [32: 244] which is "the type of use and the way boundaries are psychologically managed and identify the conditions under which flexibility promotes positive work–life outcomes". The authors posit that the way workers use and experience flexible work arrangements is dependent on how they prefer to manage the boundaries between work and home. Thus, the integration of work and life domains is not always the preferred approach and individuals may prefer to segment to some extent work and family.

In this regard, for instance, the study by Santos [47] conducted with Portuguese academics outlines the importance of the life-cycle in boundary-crossing processes. For several academics in this study, parenthood is the life domain that influences the most the crossing of boundaries between work and family on a daily basis. Thus, the existence of young children at home help in forming more impermeable boundaries between work and home because children's needs tailor the workday, while guarding against work interferences. Thus, various work factors and individual life circumstances may dictate more segmented or integrated boundaries between work and family life [5].

Theorizing FFPs: Definitions and Perspectives

Albrecht [1] defines the concept of FFPs as a set of corporate programs or measures designed to respond to employees concerns with their family responsibilities, aiming to help them manage their professional demands and personal and family needs, thus contributing to overall work and life quality. Other authors [51] characterize the FFPs as measures aimed at alleviating the conflict or negative spillover between work and family life. More recently, the designation of FFPs has been replaced by the term "work–life" initiatives [10, 12, 28]. This is a more encompassing label because it entails other life domains that go beyond the family and that include leisure and voluntary and community work. Nevertheless, our study is particularly centered on the issue of family activities and needs, as such we will be employing the more strict designation of FFPs. Examples of such measures may include job sharing, flexible scheduling, on site-day care, elder/childcare provisions, and condensed work weeks [51]. Overall, the FFPs have been categorized into three main areas: (1) measures aimed at responding to family needs, such as childcare or eldercare assistance; (2) measures regarding temporal and spatial flexibility, such as telework or flextime; (3) and, finally, legal measures, such as the granting of parental leaves [12, 21, 26].

Regarding the reasons why organizations favor the adoption of FFPs, we find distinct explanations in the literature. The most prevalent explanation relates to institutional theory [15] that posits that organizational fields become increasingly alike because there are a set of forces that pushes them towards isomorphism: "which is a constraining process that forces one unit in a population to resemble other units that face the same set of environmental conditions" [15: 149]. Three isomorphic mechanisms were detected by DiMaggio and Powell: *coercive* isomorphism relating to legal and political influences; *mimetic* processes that occur when one organizational unit imitates others units' behaviors and successful practices and policies; and, finally, *normative* isomorphism that steams from professionalization processes within the organizational field. Thus, various authors highlight the role played by institutional pressures for explaining the adoption of FFPs by organizations within a given organizational field [e.g., 7, 25, 42, 43, 51, 55].

The recent study by Pasamar and Valle [43] with Spanish HR managers in two industries—food and beverage and IT industries—shows a mixed support for both the institutional perspective and expected efficiency gains from a managerial perspective (i.e., the perception by managers of the internal benefits and costs from adopting work–life initiatives). The results show that mimetic and particular normative pressures influence the adoption of work–life practices, while coercive pressures (e.g., the strength of legal arrangements) are less effective than expected. The authors conclude that Spanish legislation is not enough to promote effectively the use of work–life practices because it might be difficult to legislate workplace culture changes [43: 1145]. Moreover, according to Pasamar and Valle, whenever the organizational decision makers perceive positive outcomes related to the adoption of work–life practices, they respond more positively to coercive and particular normative pressures, which increases the availability and use of those benefits.

Consequently, important factors in the adoption of FFPs, at the organizational and managerial levels, may refer to the size of the firm; the work force composition in terms of gender distribution and academic qualifications levels; labor shortages in specific work specialties; human resources recruitment and retention strategies; and the prevalence of management systems focused on assuring employees' organizational commitment through the benefiting of FFPs, as well as other organizational efficiency gains, such as reducing absenteeism and turnover costs [7, 9, 14, 23, 25, 33, 34, 42, 45, 55, 56]. For instance, the recent study conducted by Kotey and Sharma [33] with Australian small and medium enterprises (SMEs) points to the prevalence of low levels of flexible work arrangements (FWAs). Overall, SMEs allowed for the flexible use of leave entitlements and flexible working hours but seldom conceded the use of paid parental leave or job sharing opportunities due to their limited resources and firm size. There were also differences in terms of business sectors with the use of FWAs more common in the IT or scientific services industries than in the manufacturing or primary sectors. And this happened because, due to job requirements, the employees had to be physically present to perform certain services or tasks, which did not allowed them to work from home (e.g., in the IT sector it was more common to work from home because of the use of the internet). In addition, unskilled workers by comparison with skilled workers used much less FWAs. The rationale for that was that skilled workers held more control and autonomy over their work and they were in a position to negotiate better employment conditions and terms.

Overall, in recent decades, the research in this area [e.g., 11, 13, 29, 34, 36, 57, 59] recognizes the importance of implementing and adopting FFPs for both employees' well-being and organizational effectiveness and improved economic performance, in particular, because the adoption of FFPs helps in the strengthening of employee commitment, human talent retention, and decreased absenteeism and turnover intentions.

The Adoption of FFPs: The Role Played by Organizational Culture

Despite the individual and organizational positive consequences connected to the use of FFPs, some barriers have been detected in what concerns its effective use inside organizations. Thompson et al. [54] suggested that the lack of use of FFPs by employees may be explained by the level of support that they encounter in their workplaces, arguing that the level of work–family conflict that they may endure is more influenced by the organizational support encountered than by the availability of FFPs. In other words, in order for the FFPs to have a positive effect in reducing work–family conflict levels, it is necessary that individuals feel that their use is supported by the organization and does not put their career opportunities at risk. Thus, how the organizational culture favors (or not) the use of FFPs is a fundamental aspect in the adoption of FFPs [e.g., 3, 4, 6, 12, 28, 38, 39, 54].

Thompson et al. [54: 134] define a supportive work–family culture "as the shared assumptions, beliefs and values regarding the extent to which an organization supports and values the integration of employees work and family lives." The authors underline three important aspects that may make an organizational culture more or less supportive of the work–life relationship: (1) the expectations that the employee will prioritize work demands above family needs, which is reflected in the number of hours that is expected that the employees will devote to the organization as well as their unrestricted availability. This is characterized by Daverth et al. [12] as organizational time expectations of spending long hours visibly present; (2) the perceived negative career consequences or outcomes of adopting the available FFPs, [12]; (3) and the perceived organizational support and sensitivity to employee's family responsibilities and needs. This is referred by Daverth and colleagues [12: 1712] as the managerial and organizational support for the uptake of FFPs. These authors include two more determinants of a supportive work–family culture: (4) the gendered perception of policy use that is the prevalent idea that FFPs pertain only to women; (5) and, lastly, co-worker support regarding work–family issues. Thus, individuals are more likely to benefit from the use of the FFPs in a supportive organizational culture, thus experiencing reduced levels of work to family conflict [54].

One of the most studied organizational barriers refers to the cultural stereotype of the "ideal worker," which contributes to the maintenance of workplace inequalities. The "ideal worker" is someone that displays an undivided time commitment and loyalty to the organization and is not distracted by family responsibilities, being often depicted as a male worker [34]. This idea of the "ideal worker" is deeply rooted in most working organizations that assume a preference for the employees segmentation of work and family life domains, and which usually results in a greater appreciation of male workers or of those workers without major family commitments. In most organizations success and commitment is thus equated with working long hours—which is usually more problematic for women that are still the primary caretakers of the family—, and those who contest this culture and exhibit their family commitments are perceived as less organizationally committed [8, 10, 22, 28, 34, 48].

Although there is an effort to make the issue of work–family interaction neutral from a gender point of view, in practice, FFPs tend to be seen by both the employers and the employees as policies broadly directed towards women [14, 23, 34]. According to Gregory and Milner [28] men often benefit from some type of informal work flexibility but in a more concealed way and not explicitly connected to family needs. Thus, FFPs are stereotyped as 'women policies' reinforcing more than questioning gender stereotypes [34] as women continue to be perceived as the main caretakers of the family and not committed enough to their organizations.

The study by den Dulk and de Ruijter [14] indicates that managers are essentially interested in the performance of their departments and that work–life policies are often seen as disruptive of work tasks and performance goals. In this research, manager's decisions to grant FFPs were dependent on whether the employee making the request was supervising others, which was considered less favorably, as well as on the gender of the employee. Requests made by women were considered more favorably than those made by men, in particular when it concerned taking up parental or care leaves. According to the authors, this is an indicator that family duties are seen mainly as the responsibilities of women, which might curtail men's opportunities of also undertaking FFPs. Likewise, in the Greek context, Giannikis and Mihail [23], pinpoint that the adoption of flexible work arrangements was mainly uptake by women and public sector employees who were more likely to perceive more benefits and fewer costs concerning the use of work flexibility.

The adversarial organizational culture is also reinforced by the lack of supervisory support. McGowan [38] found in her study that although managers publicly defended that the employees with responsibilities in caring for the elderly should present their cases to management, in practice they advised them to keep family matters in the private realm. According to the author, the paradox between the recommendation given by managers to explicitly express their family needs and, informally, their advice to keep family issues silent highlights the power of organizational cultures to resist change and inhibit the use of FFPs. In this particular case, the following of a silence strategy, by not communicating personal matters, shows that the employee understood the (informal) message conveyed by the organization about the proper ways to manage family commitments, emphasizing the individual strategies, in which silence is included to the detriment of formal organizational policies [38].

Hence, the support of supervisors or immediate managers is crucial for employees' successful work–life integration [12, 13, 20, 30, 35, 36, 46, 50]. As underlined by Hopkins [30: 449] supervisors act as organizational gatekeepers regarding "worker's knowledge and use of organizational benefits, resources and programs that might help workers better manage work and life responsibilities." Additionally, workers trust that their supervisors will help them manage their work–family responsibilities by being knowledgeable about the family-responsive policies or benefits available to them. The recent study by Daverth et al. [12] proves that the existence of a "context of support" is essential in the uptake of FFPs by employees and that middle managers tend to mirror the behavior of senior managers relatively to the support that they extend to the available work–life policies, thus keeping with their perceptions of senior management expectations. The interdependence between organizational culture and structure (i.e., formal HR policy and practice relatively to FFPs) was then fundamental. Moreover, the study by McCarthy et al. [36] shows that both organizational support (i.e., from the HR manager) as well as the immediate supervisor support affect employee uptake of work–life programs. In this study, employees reported more positive work–life outcomes such as less role conflict, greater job satisfaction, and less turnover intentions, when they perceived their supervisor as supportive.

Additionally, the support obtained from co-workers is also important because it may enhance or inhibit formal initiatives aimed at managing work and family [2, 12, 13, 31, 40, 53] with research findings suggesting that co-workers may play a vital role in the work–family relationship by providing emotional and/or instrumental support. Emotional support is related to the display of respect for the personal and family problems of work colleagues, while the instrumental support refers to the completion of co-workers tasks whenever needed. The authors found that emotional and instrumental support were more likely to occur in cohesive groups, where employees considered the work climate as family-friendly, and believed that the procedures relatively to the use of FFPs were managed fairly, and recognized that their own supervisors supported the solving of conflicts between work and family. Mesmer-Magnus and Glew [40] further note that the messages coming from supervisors, consistently reinforcing the importance of family support, have a positive impact on co-workers behaviors and attitudes in terms of also reinforcing a supportive work–family attitude amongst them. Also, the study by de Sivatte and Guadamillas [13] shows that co-worker utilization of flexible work arrangements was one of the strongest predictors for the effective use and adoption of various flexibility benefits by employees (e.g., flextime, extra vacation days, or telework).

An important implication of the studies above is that it is not enough to create FFPs, as these have to be supported by immediate supervisors and managed with fairness throughout the organization and among distinct categories of employees. To sum-up, the improvement of human resources management practices should consider the analysis of the symbolic and cultural dimensions of organizations, thus further advancing organizational work–family supportiveness.

Methodology

Research Design and Study Participants

To carry out the empirical study, a research strategy guided by a qualitative approach was adopted with the use of the case study method [19, 60]. According to Eriksson and Kovalainen [19] the case study design offers a detailed and holistic knowledge on the subject under-research because it is based on the analysis of multiple empirical sources of data, which are rich in real context. We employ an extensive case study design by using four companies. Taking in consideration our research questions, the study goals were the following: (1) to identify the FFPs that four companies made available to their employees; (2) to isolate differences or similarities between them relatively to the work–family organizational culture, namely the role played by the immediate supervisor and co-workers in the adoption of the FFPs.

The case selection was based on theoretical sampling aspects [19] and a convenience sampling strategy was used. Thus, by selecting these four cases we expected to find similarities as well as differences between them that might allow us to verify existing theory and/or make comparisons. The sampling criteria were the following: (1) to have companies belonging to different business sectors or industries in order to make comparisons and find differences or similarities between cases; (2) to have companies with a human resource management (HRM) department and with different sizes because this would enhance the variety in what concerns the availability of FFPs. Table 1 offers a detailed description of the study participants.

For reasons of confidentiality and anonymity, the names of the companies are fictitious—hereafter designated as Alfa, Beta, Delta, and Chi. For each company we interviewed the HR manager as well as five other employees, totalizing 24 participants. Regarding the marital status and the number of children, most interviewees were married or cohabiting and a few were divorced or single. Eleven

Table 1 Socio-demographic characterization of the interviewees

	N = 24	%
Gender		
Men	8	33.33
Women	16	66.67
Age		
Under 30 years	2	8.33
31–45 years	19	79.17
46–60 years	3	12.5
Educational level		
Preparatory level	3	12.5
Secondary level	10	41.67
Graduate level	11	45.83
Hierarchical position		
HR manager	4	16.67
Supervisor/manager	6	25
Operational	14	58.33

interviewees had 1 child, another 11 participants had 2 children, and only 1 interviewee had more than 3 children, whereas another 1 had no children. In most situations, the children were under the age of 12 and living in the family household.

Data Collection and Analysis

To collect data the following instruments were used: various organizational documents (e.g., organizational mission statements, annual reports, and company websites), direct observation and field notes, and qualitative interviews. The gathering of documents was useful for a general characterization of the company and its human resources strategy. The access to the empirical data involved the making of several visits to the companies, which provided the possibility of direct observation and field notes, thus registering some observations and comments either during the visits or immediately after them.

The main instrument of data gathering was the semi-structured interview which involved the outline of main topics, by using an interview guide, but that differed according to the interviewee reasoning [19]. Two different interview guides were developed, one aiming the responsible for HR management and the supervisors and the other one intending the operational employees. The length of each interview comprehended a minimum of 30 min and a maximum of 90 min and they were all recorded and fully transcribed. The interviews were conducted during the year of 2013, and the majority was held at the company facilities with the exception of two interviews that were conducted outside the company at the interviewees' request.

The data gathered through the interviews was subjected to a theme or coding analysis [24]. Hence, some codes were defined previously based on the literature, whereas other codes were established during the data analysis. Given that two different interview guides were drawn up, one for the HR managers and other for the employees, we developed two grids of analysis. The findings revolve around four main themes: (1) types of FFPs that are available to employees; (2) HR managers and employees perceptions about the FFPs; (3) industry or business characteristics that facilitate or hinder the adoption of FFPs; (4) and, finally, characteristics of the work–family organizational support culture.

Findings

Case 1: Alfa Company

The Alfa Company was established in the 1980s and is part of a Portuguese multinational corporation that operates in the food retail industry. The company is geographically distributed into two distinct locations, and employs an overall number of 1013 employees of which a significant number are women. The Alfa Company has always supported employees in terms of health insurance benefits or various financial assistance programs through the establishment of an internal corporate social responsibility (CSR) department that functions as a support unit for employees. The CSR department unit is nowadays part of the company formal structure, having its own human and financial resources and an annual action plan that is built around employees needs and upon consulting them. The CSR department possesses the explicit support of top management and its initiatives are closely linked to the top administrator leadership style that is highly paternalistic. For a detailed description of the main FFPs and company benefits of Alfa Company, see Table 2.

The corporate initiatives were conceded to all employees regardless of the function, the type of contractual arrangement or job seniority but its concession was conditional to the employee's annual income (e.g., the education and social assistance initiatives were especially designed to benefit employees with lower incomes).

The following statement from an Alpha employee is illustrative of the paternalistic-oriented work–family policies of the company:

> When my wife got pregnant we looked for day care and kindergarten prices, (…) and the minimum monthly price was 150€… So, each month I manage to save 150€. And I see that assistance benefit as a salary extra, right? If I have a monthly salary of 550€, and if I have my child kindergarten for free, I feel that I earn 700€! (Man, operational worker, 39 years)

In the case of the highly educated employees, holding top-level management positions, the Alfa Company offered other benefits, such as a personal laptop and a mobile phone to have remote access to the workplace, and a health insurance plan that was extended to all family members. The HR manager in this company concurred with the perspective that the work–life relationship should be viewed as integrated and not as segmented:

Table 2 Corporate initiatives and benefits implemented by the CSR department of Alfa Company

Corporate programs	Type of initiatives or benefits
Health and employee welfare	Health insurance internal program (free medical consultation or reimbursement of 90% of medical expenses)
	Wellness week
	Recreational and sports activities
	Gym activities and physiotherapy
Education	Support of employees' academic and educational goals (e.g., allowing them to pursue the completion of a secondary school degree during working hours)
	Granting of scholarships to employees and/or their children
	School kits for employees' children who start elementary education
	Discounts in the purchase of textbooks, school supplies and clothing
	Full payment of school books to large families
	Summer camps for employees' children
Social assistance	Financial support (e.g., grants) in extreme situations, such as a prolonged illness, death of a close family member, and exceptional financial assistance)
	Baby's kits upon birth
	Life insurance
	Telephone helpline to employees providing assistance in legal matters or other rights clarifications
	The offer of family Christmas vouchers
	At one of the company locations, there is a nursery and a kindergarten that is free for employees' children

> In the past there was this mistaken belief, which annoyed me that 'work was work and family was family and that they should be kept separately'. Fortunately, our top managers understood it differently as time passed and that has changed. (…) Boundaries in life get blurred… (Man, HR manager, 43 years)

The work schedules in Alfa Company are continuous, established in a 24 h and 7 days week basis with the use of shifts. Hence, continuous work schedules bring about some difficulties in terms of family life management since most of the schedules do not have entry and exit times coincidental with the schedules from other institutions such as schools, day care centers and study centers, or even public transport, and the schedules of other household members. The company offers flexible work arrangements but acknowledges that the nature of the business and company performance goals do not always match the personal needs of all employees, especially operational workers that have to choose between shifts. Here is an excerpt that illustrates this point of view:

> In our line of work we have the stores that are open daily during 14-16 hours a day, thus we have multiple shifts with different work schedules…an opening shift and a closing one, and so on. In terms of logistics, its non-stop, the work schedule is 24 hours a day and 7 days a week. Hence, if the company wishes to meet its performance goals and remain competitive,

something is got to give, and usually it is family and personal life… (Man, HR manager, 43 years).

Interestingly, some of the operational workers in Alfa Company were resorting to individual strategies as a way to manage their work and family lives. Hence, the interviewees mention the use of informal support networks such as neighbors, babysitters or other family members. The company has a significant number of couples that are both employees of the company, and sometimes they try to juggle their work schedules in order to take turns in the fulfillment of family responsibilities, thus sacrificing their living as a couple. The sacrifice of time to rest and sleep was also frequently mentioned.

When negotiating with their immediate supervisors, many operational workers try to bargain an informal solution which, in most cases, refers to schedule flexibility that allows them to make switches in terms of entry and exit times, days off, number of hours worked per day or vacation periods. In this context, the fact that there are continuous working hours is considered an advantage from the perspective of employees since the margin for exchanges with other co-workers is very large. However, this measure is used informally and at the discretion of each supervisor, which seems to pose problems in terms of fairness between distinct employees. Thus, some workers report situations of support and comprehension for family matters by their immediate supervisors and of mutual support:

My supervisor has always trusted me and she has always been very flexible regarding my family and personal needs overtime. If I tell her: 'tomorrow I'm not able to come because I have to do this and this…' and even if she does not have anyone available to replace me, it has happened before, she would replace me herself… and I have also done that for her. There were situations in the past where I had to cover for her. (Man, operational worker, 34 years)

By comparison, other participants convey situations of preferential treatment between employees and lack of support and understanding on the part of their supervisors, pointing to a scenario where productivity interests and compliance to organizational goals prevail exclusively:

I think the company should pay more attention to those who have young children at home as well as elderly family members, because it is complicated in those situations. Allow us with more flextime and more freedom in terms of schedules because we often need to go to the doctor's and sometimes we need an hour or two and they [the supervisors] don't let us, they say no! But when they need us to do overtime, we are always amicable to their requests and we receive them with open arms! (Woman, operational worker, 30 years)

Statements were also collected that refer not only to barriers posed by immediate supervisors but also discouragement to seek support at other levels of the organization, such as the HRM department:

I know that I had the right to a schedule change, to make an adjustment. I know that if I wanted to have it my way, I should have complained to the HRM (…) and I think that I would have wan. But, should I do that? Start a war with my supervisor? Go behind his back and complain to the HRM? No, I don't think so… (Woman, operational worker, 37 years).

It seems from the testimonies that some employees do not feel at ease to address their concerns at the HR level because they believe that this puts into question the legitimacy of their immediate supervisors and they may fear paybacks for confronting their decisions. Explanatory reasons for the diversity of situations encountered regarding supervisor support may be related to the size of the company and the complexity of its hierarchical structure, with four hierarchical levels within a single unit or store, and the lack of formal policies regarding the concession of various flexible work arrangements.

As regard to the support provided by co-workers, the statements of the workers report evidences of emotional and instrumental support, with colleagues available for shift switches among them and showing concern for the well-being of their colleagues. Some statements also show that co-workers can act as a strong cohesive group and with power to alter dominant management practices, including the imposition of limits on overtime requests by their supervisors. Overall, most operational workers interviewed at Alfa Company seemed to be unaware of the availability of concrete work–family policies despite the company formally displaying a broad array of CSR programs and benefits (see Table 2).

Case 2: Beta Company

The Beta Company is a Portuguese multinational company founded in the early twentieth century that exerts its activity in the industrial sector and has 1092 employees working in Portugal. Beta Company workforce is mostly male—90% of the job positions. Additionally, the company possesses seven factories, of which three are operating in Portugal and all of them are run in a highly automatized way with cutting-edge technologies. Thus, the HR manager underlines the difficulty in recruiting personnel for highly specialized technical functions at the operational level as well as for some management positions that require skills that are scarce in the labor market.

In 2011, Beta Company started the formal process of obtaining social responsibility certification. The implementation of the social responsibility norm was followed by several initiatives aiming at sensitizing the employees to social responsibility matters. The company is also known for its strong links to the community and its sponsorship of several local initiatives. Although the company recognizes the importance of the work–family relationship for employees' job satisfaction, with the need for work–life conciliation referred to and endorsed in institutional communication letters, the HR department recognizes the absence of formal FFPs:

> We do benchmarking by comparing ourselves with other companies that operate in the same sector, that's how I know about this subject [work- life policies] and its importance nowadays (…) I would say that in terms of formal procedures, hum… with don't have any, no. Of course, if there is a situation that for some reason, an employee has a serious family

problem that requires an exceptional measure… we analyze it, this has already happened. But a formal procedure (…) we have nothing established. (Woman, HR manager, 40 years)

Despite this, Beta Company offers a set of measures or work–life benefits that are, nevertheless, differentiated according to the employees' functional level and rank (operational *versus* managerial and administrative functions) and that include:

(1) A family open day: it is a day in the year where employees are allowed to bring their family members to the workplace to know the factory facilities. This initiative was also extended to employees' friends in the current year.
(2) A Christmas party for the employees' children with gift giving.
(3) Flexible work arrangements but only for administrative personnel and managerial levels.
(4) The granting of 23 days off to the operational workers that work on rotating shifts.
(5) The possibility to work remotely from home that is offered to managerial employees on special situations.
(6) The inclusion of issues related to work–family conciliation on career development plans for employees in managerial positions (e.g., the allocation of geographic moves or expatriate assignments).
(7) The granting of a health insurance that is extended to family members, but only for employees at managerial positions, and the concession of favorable conditions of adhesion to the health insurance for all other categories of employees (namely, operational workers).

The absence of formal FFPs, particularly at the operational level, is explained by the HR manager in two ways: (1) on the one hand, work is allocated according to rotating shifts, which introduces some rigidity in work schedules and makes it difficult to implement flexible work arrangements, such as shift exchanges or changes in terms of hours of entry and exit (they operate with a morning, afternoon, and night shift and every worker rotates between shifts for given periods of time); (2) on the other hand, the employees do not seem to experience difficulties in terms of balancing work and family life, and according to the HR manager, they have the opportunity to do so through the work climate questionnaire that is administered biannually to all employees.

In addition, working in shifts is seen, at the managerial level, as an advantage because there is the possibility of a clearer separation between work and life outside work, as the worker ends his shift he also leaves the work load for the next shift that comes in. On the other hand, the possibility of granting flexible work arrangements is higher in the case of the employees holding managerial functions. The following testimony is illustrative:

Operational workers, at the factory, also have family needs and problems to resolve, but as they work in shifts it is easier to deal with those at the end of the shift, whereas I cannot simply leave my workplace during working hours, my work schedule is always the same.

> As they work in rotating shifts, they sometimes have to be here during the day, but other times they don't. (…) In my case, what I often feel is that if I missed a working morning because of a family matter, then I have to compensate the company, and I take the work with me to do at home. (Woman, managerial position, 35 years)

In the Beta Company, the 23 days off that are granted to operational workers in rotating shifts seems to be the explanatory reason for the high levels of employee satisfaction with the company policy. According to the HR manager, in the last working climate survey more than 80% of the employees signaled that they were able to balance their personal and family life with their work.

Additionally, another motive that may justify the satisfactory results is the fact that most of the workers are male. According to the HR manager, the work–family policies are more justifiable in organizations where most of the workers are female, which is an indicator of taking gender into account when thinking of FFPs, and of the pervasiveness of gender stereotypes at the decision-making levels. The fact that the division of household chores still remains much in the hands of women means that organizations with mostly female workers may display a higher probability of adopting FFPs, either for reasons of recruitment and retention or for organizational efficiency reasons, such as reducing absenteeism and/or improving productivity. Lewis [34] underlines that by centering FFPs on women, organizational decision-makers are not promoting diversity goals or gender equality but are marginalizing the issue of work–family conciliation by making it a woman's problem:

> I think that another reason why we never had a formal policy on that matter [FPPs] has to do with our workforce being male.… My personal opinion is that in the Portuguese private sphere, it is still the woman that is responsible for household chores and family care, so I believe that companies with mostly female workers are certainly confronted with issues that do not affect us much. (…) Most workers here are husbands and… I, do not see these requests: 'Oh, my child is sick, I have to take a family leave'… (Woman, HR manager, 40 years)

The HR manager describes the organizational culture has being strongly influenced by the top administrator, which has been heading the company for 20 years: "*I believe that our organizational culture has much to do with our President and our Executive Board and how they look at these things, and the messages that they convey to us about it. And what they tell us is that: 'family always comes first'.*" In this regard, the HR manager states how working overtime is not a common practice and how administrative employees are encouraged to leave the company on regular hours. In those cases, there is also the possibility to finish pending tasks through remote access to company computers. Nevertheless, in Beta Company the interviewees also show a lack of knowledge regarding the availability of organizational practices to support the conciliation of family life and work.

Case 3: Chi Company

The Chi Company is a Portuguese company with 20 years of existence that operates in the new information technologies sector. The Chi Company is a small-size enterprise with 98 employees, with a balanced distribution by gender, most employees are highly qualified personnel in the information technology and management areas. Presently, the Company is a member of the 'Portuguese Network of Corporate Social Responsible Companies' and integrates the European Guide on Corporate Social Responsibility. The Chi Company is involved in several initiatives with the local community and some of its employees do volunteer work in the community. The company is characterized by few hierarchical levels and a corporate history strongly marked by its main founder, who continues to run the company. According to the HR manager, his leadership style has always been characterized by proximity, innovation and entrepreneurship.

In terms of work–family benefits, the Chi company does not possesses any formal policies and is also characterized by policy informality on those issues, as explained by the HR manager:

> We have a set of measures, of initiatives, that are being implemented and developed but that are not formal measures, that is, we have some family- friendly practices that are in use but that are not on paper. What are they? Let's see, there is no formal policy stating that an employee has the right to skip a work's day to go with a family member, a grandparent or a child, to a doctor's appointment, but that already happens here and we have the availability to accommodate that type of request. There is, on the company side, a concern with the issue of work- family conciliation, I would say that… (Woman, HR manager, 37 years)

Despite the Chi Company being an intensive user of IT technologies, the company does not offers the possibility of adopting flexible work arrangements to its employees (there is the exception of the Research and Development departmental unit). Once again, this is explained by the HR manager has being linked to the characteristics of the work tasks that are strongly customer- service oriented (on a business schedule that runs from 09.00 am until 19.00 pm), which means that the employees have to be physically present to attend the customer, thus limiting the possibility of more flexible work schedules. Nevertheless, the HR manager present us with a set of family-friendly benefits and social responsible initiatives that are used by the company although not formalized:

(1) The possibility to have schedule flexibility or to work remotely from home but on an informal basis (this is the case of the employees working in the R&D department and that are software developers).
(2) A health insurance that may be extended to the employees family members.
(3) The celebration of some dates, such as Fathers' day and Mothers' Day.
(4) The granting of a financial bonus of 500€ to the employees that become parents.
(5) Several initiatives of community proximity that also involve the employee's family, such as doing joint volunteer work in the community.

Overall, the interviewee's testimonies show us a supportive organizational culture in terms of supervisor and co-workers support and the sensitiveness to work–family problems and difficulties. The following statement is illuminating:

> My co-workers are also women and they are very understanding of my family needs… We help each other out and we substitute for each other whenever we need (…) For instance, when scheduling my holidays the other day my colleague even said to me: 'you have your kids and your husband to worry about, so you see first which dates you prefer to book'. (…) And if I need to take an hour or two hours in the work day, I talk to my supervisor and I come in earlier in the next day. I'll make up for it another day…. (Woman, operational worker, 34 years)

Despite the testimonies of employees reporting a working climate of proximity and mutual support, and supervisor openness to family issues, some employees underlined that the effort to balance work and family life was essentially an individual effort and not something in which the company played an active role:

> I do not have any extended family nearby, so I and my wife have to manage everything without family help, so…in the days that the school closes, we already know that we have to take some vacation days in order to stay with our children. Usually, we both try to take those days off, so that we can have some family time together. (Man, operational worker, 34 years)

As with the previous companies, Chi employees demonstrate a lack of knowledge about the existence of family-friendly benefits or practices inside their organization and they display difficulty in identifying measures to support that conciliation. They usually pinpoint the importance of having schedule flexibility, considering this a subject about which they have never reflected upon.

Case 4: Delta Company

The Delta Company operates in the Portuguese hospitality sector and it owns seven units that vary between 3, 4, and 5 stars city hotels. The Delta Company was founded in 1986 and currently employs about 220 employees. The company is characterized by many hierarchical levels and the emphasis on employee productivity and organizational profit arises several times during informal conversations with employees. The Delta Company is also characterized by the lack of formal policies in the area of work and family conciliation. The HR department sees the issue of work–family conciliation as not making part of the department's main goals. In this company, the HR department has a strong administrative and legal character, and is essentially focused on administrative tasks, such as recruitment and selection procedures and salaries and compensation issues. The areas of training and internal communication are clearly underdeveloped. The lack of interest by the top administrators about work–family issues is conveyed by the HR manager interviewed:

> In the short term, the main company stakeholders [shareholders and top administrators] do not recognize that the implementation and development of such policies and practices

> [FFPs] will carry a significant weight on the return of financial investments and, therefore, they are not interested in making any investment. I do not even know if they ever thought about this… I guess they never sat down to think about it. (Man, HR manager, 43 years)

Some of the interviewees at Delta Company clarify that the lack of organizational initiatives concerning work and family conciliation may be explained by a distant Administration that emphasizes financial short-term results, and that uses aggressive strategies to reduce personnel costs, a management style which has become more pronounced in the recent four years due to the severe economic crisis that affected the country between the years of 2011–2013:

> I think that the company is not suited to deal with issues of work- family conciliation, because that also entails some financial costs, and maybe they do not have that money available for that, and as long as they pay the salary to the employees they think that they are okay, that the company is fulfilling their part. (Woman, managerial position, 50 years)

In the last years, Delta Company introduced several changes that were depicted negatively by the interviewees and that led to the augmenting of employees' unionization. Among those dissatisfactory measures, the company decided to cut back on food subsidies and replace them by meals and, at the same time, demanded from employees' greater geographical, schedule, and functional flexibility. The HR manager states that the employees may refrain to make more requests regarding flexible work arrangements because, on the one hand, they may not feel entitled or may feel unease making the requests to their supervisors and, on the other hand, they may be afraid of being let go and lose their job. The use of flexible work arrangements, in the case of Delta Company, seems to be one-sided and work only to the company advantage. In Portugal, the hospitality industry is characterized by short-term contractual arrangements. Here is an excerpt taken from an interview with the HR manager that corroborates this rationale:

> Some people do not want to ask for schedule flexibility because having worked in the company for a long time, they are aware of the workplace culture and the need to show a great availability for work and do overtime. And also because they think they do not have the right to ask (…) Here in this company, the issue of schedule flexibility is critical, it is a critical point, the employees who do not show the openness in terms of working long hours and be flexible, they may soon be 'invited' to leave. (Man, HR manager, 43 years)

At times of economic crisis and financial strain, some organizations may reduce the benefits granted to employees in order to remain financially viable [10]. This might be the case of Delta Company that seems to consider the issue of work–family as a strictly personal matter and that considers that assuring the employee with a monthly salary is what the contractual arrangement obliges. This company positioning is reinforced by the current Portuguese socioeconomic context of high unemployment rates and welfare retrenchment [61]. The following statement depicts well how employees feel about the company top management:

> Our managers are not concerned with that, there are no such things as family- friendly policies! Our managers' mentality is very short-term and one-sided. They should know that we are human beings and that we may get sick or have other personal stuff to manage, we are people, not machines. (Woman, operational worker, 47 years)

Once more, the role played by immediate supervisors seems to be crucial. At Delta Company some employees acknowledge the support given by their supervisors and co-workers in family related matters, who are often available to facilitate schedule changes and assist in completing tasks. This support seems to be an informal practice that may vary according to the supervisor management style. In the absence of written policies, immediate supervisors are empowered to authorize schedule flexibility, which translates into the possibility of the employee changing entry and exit time or the number of hours worked in a day. Given this supervisory autonomy and informality, some references to unfair situations appear, as shown in the following excerpt:

> It is obvious that sometimes there are two weights and two measures. In a section of 10 people, they are not all treated in the same way, and there is the chance that a given person who was ill of staying home and make up for it later, while another person who was also ill got her salary cut. (…) With the economic crisis, people hold back, do not speak of injustices and do not express this feeling. (Woman, operational worker, 47 years)

Conclusions

The current study aimed answering two research questions: (1) What type of FFPs can be currently found in employing organizations?; (2) What are employee perceptions about the organizational support regarding the various FFPs that are made available to them by their organizations?

Regarding the first question, it was clear that in both four case study units there were no formal policies or practices regarding the issue of work–family conciliation. The HR managers interviewed in the four companies have a general knowledge of the subject under study, but acknowledge the absence of formal work–family conciliation policies and practices in their organizations. Only in the case of Alfa and Beta companies was there a more developed set of work-–life benefits that were, nevertheless, part of the CSR strategy being followed, especially in the case of Alfa Company. More importantly, in both four units what seems to prevail are informal policies regarding flexible work arrangements or small adjustments in terms of daily work hours or schedules (e.g., shift exchanges). Thus, like happens in Townsend et al. [56: 16] study, our findings show that what prevails is an informal system of practices that is associated mostly with the "utilization of ad hoc, short-term flexible arrangements" that are granted by immediate supervisors. Thus, the ad hoc solutions shift the work–family responsibility away from the organization and to the individual [28] and hinder an effective organizational change, often contributing to the maintenance of unfair management practices that are legitimized by the labor market and organizational financial needs.

Additionally, the findings also reveal that the prevailing perspective in terms of HRM is a legalistic one, since organizations are only concerned with complying with parental leaves and with the granting of other leaves allowed by law (e.g., sick leaves and legal protection of pregnant or postpartum women workers). For instance, no company had formal arrangements related to the care of elderly.

Like in the study by Pasamar and Valle [43], we found that institutional pressures [15] influenced those companies adoption of work–life benefits especially due to normative and mimetic pressures. For instance, the HR manager interviewed at Beta Company clearly stated that the company benchmarked others in the industry in order to know the 'best practices.' This is one reason why the company offered, for example, a health insurance, which was a common practice in the organizational field (mimetic pressure). Corporate certification in terms of Social Responsibility (normative pressure) awarded the companies with added legitimacy in their respective organizational fields as well as the reputation as a 'social responsible company.' Nevertheless, there were other reasons, related to organizational efficiency gains and managerial reasons that also explain why the companies in our case study opted not to offer FFPs. Thus, very alike the findings in Pasamar and Valle study [43], we have to consider not only external pressures from the organizational field but also the assessment being made, by organizational decision-makers, concerning the internal benefits and costs of adopting work–family-friendly policies and practices, which clearly seemed to be the case of Delta Company.

In times of economic crisis, organizations may refrain from adopting FFPs because of financial constraints, which might be the case of some companies in the current Portuguese context due to a prolonged economic crisis that reached its peak in the period 2011–2013 with an international financial assistance program that avoided State bankruptcy [61]. Hence, it might be the case that managers in Delta Company are able to "distance themselves to some extent from being accountable for decision making in response to request for flexibility through reference to externally derived financial constraints" [56: 18].

Other important organizational characteristics that should be considered when adopting FFPs are organizational size and workforce characteristics [33, 56]. In our study, the large size companies (Alfa and Beta) offer more work–life benefits overall and this might be explained by the availability of more economic resources. At the same time, workforce composition seems to influence the adoption of FFPs. For instance, in Beta Company, the HR manager interviewed considers that implementing FFPs would make more sense if the company had more female workers, because family care is mainly a women's responsibility, thus being the company workforce predominantly male there is no need for those policies. This shows a gendered organizational culture, which has also been reported elsewhere [e.g., 23, 34, 48], and that ends up reproducing gender inequalities inside organizations.

One distinctive aspect of our findings is that within each case unit, with the exception of Delta Company, there are differences between the various categories of employees regarding the type of work–life benefits available to them. For example, in Beta Company only the managerial and administrative employees could benefit from remote work or flexible work arrangements, which were not granted to operational workers. In Chi Company, only the software developers that belonged to the R&D unit were allowed to work from home or to have irregular work hours. Thus, the nature and the type of work activities or tasks entails differentiations in terms of the work–family measures that might be granted to each category of workers. This has also the potential of generating perceptions of unfair treatment amongst the different categories of workers within each company. See Table 3 for a summary and comparison of the main organizational characteristics of the four case study units.

Table 3 Comparison of the main organizational characteristics of the case study units

	Companies: summary and comparison			
	Alfa	Beta	Chi	Delta
Company size	Large	Large	Small	Medium
Geographical dispersion of the business	Large	Medium	Small	Medium
Type of operational activities	Routine activities but highly dependent on influxes from the external environment	Highly automated activities, routine but highly specialized tasks	Non-routine tasks R&D activities	Routine activities but highly dependent on influxes from the external environment
	Not specialized		Customer-service oriented activities	Customer-service oriented activities
Work–family (informal) policies and practices	Some, differentiated by job function	Some, differentiated by job function	Few, differentiated by job function	Nonexistent
Schedules	Various and diversified	Rotating shifts at operational level	Regular work hours	Various and diversified
		Regular work hours		
Top management leadership style	Supportive	Supportive	Supportive	Unsupportive
	Paternalistic promotion of employee loyalty and commitment	Promotion of employee loyalty	Emphasis in employee participation and development	Instrumental view of the employee, focused on organizational profit, and short-term financial outcomes
	Explicit support to HRM	Explicit support to HRM	Explicit support to HRM	Lack of support to HRM
Supervisor management style	Inconsistent: supportive or uncooperative	Supportive	Supportive	Inconsistent: supportive or uncooperative
	Often resistant to the implementation of FFPs			
Co-workers support	Instrumental and emotional support	Instrumental and emotional support	Instrumental and emotional support	Instrumental and emotional support

Hence, those employees at the operational level that performed routine tasks and that were under a tight control due to the technology in use or that occupied a client—service function had less access to FFPs compared to more qualified categories of employees who had more access to flexible work arrangements used as a way of retaining scarce talent. This finding is consistent with other studies that point to the fact that the possibility of flexible work arrangements is dependent on the possibility of task reorganization and workers replacement by co-workers [59], as well as on the scarcity of the worker's skills in the labor market [49].

Moreover, at an operational level, in the Alfa and Delta Companies the use of informal flexible work arrangements worked mostly to the advantage of the company and the need to adjust workforce volume to work loads and business seasonality, which was resolved through the implementation of banks of hours that covered, above all, the company seasonal needs. Additionally, working in shifts or in diversified schedules, is considered by some interviewees of the Alfa, Beta, and Delta Companies as a practice that helps them to better conciliate work and family life because they are able to change schedules or days off with co-workers, allowing them to accompany their family more regularly or deal with personal matters.

Regarding the second question included in our study on employees' perceptions about the organizational support regarding the various FFPs that were made available to them by their organizations, our findings highlight that most of the employees were not able to identify the availability of FFPs inside their organizations and were mostly unware of its existence. With the exception of the Delta Company, all the other companies offered a set of health, education, and leisure benefits that sometimes entailed the establishment of protocols with external entities. Some of these benefits could be extended to other family members, such as the health insurance. The lack of awareness on employees' part may mean that internal communication processes are not the most effective or even that the management is not interested in largely publicizing those benefits.

Some studies reveal that the belief that work–family conciliation is a private matter may lead employees to hold a low expectation of support from the employer, which is legitimized by the low probability of employee mobilization to access their rights [41]. Thus, individuals may seek family support essentially from informal networks (such as neighbors, friends, or extended family) and by acquiring childcare assistance through nannies and kindergartens [52]. Thus, in our study the employees were resorting to individual coping strategies by using the support of the extended family or nannies. Additionally, balancing work and family life was also possible at the expense of sacrificing time to rest: "*In average I sleep 4 h a night. (…). And when I see that I won't have time to organize everything at home, I don't sleep*." *(Woman, 42 years, Alfa Company)*

Additionally, the organizational culture of support towards the work–family life relationship proved to be stronger in organizations like Alfa, Beta, and Chi, where top management seemed committed to the issue, not only verbally, but also through its inclusion in the overall CSR strategy (e.g., see Table 2 for Alfa Company). This finding concurs with several other studies that underline the importance of top management in promoting a culture of support [21, 37, 58].

Also important in this study is the support of the immediate supervisors in the granting of access to flexible work arrangements on an informal basis, thus confirming the evidences brought by other studies that also pinpoint the relevance of immediate supervisor support for the uptake of work–life programs by the employees [12, 13, 30, 36, 50]. Hence, the organizational culture of family support is created and sustained by the behaviors, attitudes and values communicated not only by top management, but also by immediate supervisors and co-workers, who, as a collective, weave the workplace culture. Thus, supervisor support play a crucial role as they are responsible for allowing the adoption of flexible work arrangements by employees.

Several statements made by the interviewees in Alfa, Beta and Chi companies show how the understanding attitudes of immediate supervisors and co-workers were important sources of emotional and instrumental support, hence contributing more than formal support to the work–family life conciliation. The few studies that depict co-worker support confirm that it is a type of support that increases job satisfaction, performance, and employee well-being and that reduces conflict levels [16]. In this study, in all the companies the support from co-workers was frequently acknowledged as relevant.

One important aspect that should be underlined is the supervisor or manager discretion concerning the management of informal flexible work arrangements, which may be a source of potential inequalities between workers. The availability of flexible work arrangements is internalized as part of a 'give and take' in the relationship between employer and employee. However, as Lewis stated [34], when the conciliation policies are designed as a reward to be earned, workers exhibiting less availability for the company feel a lack of legitimacy to use them. On the other hand, the fact that flexibility is a reward for an additional allocation of time, creates a paradoxical situation where people less able to give their free time to work are in disadvantage to benefit from these policies that have been implemented to help them [34].

This research has some limitations that should be discussed. First, it is not possible to generalize the findings to other contexts. Nevertheless, the purpose of an exploratory study that is qualitative is not to statistically generalize its findings, but rather to enable theoretical replicability, which this case study makes possible. Second, the study participants in each case unit were selected by the human resource managers, which might have added the effect of social desirability to their answers. Nevertheless, there was the concern by the researcher to assure the full confidentiality and anonymity regarding the information being provided by them. Also, the study participants were limited, in most situations, to those having children. The option to include participants with other family configurations might lead to different and/or complementary information on the subject under study and a more complete picture of the phenomena.

Future research suggestions include: (1) a deeper analysis about the differences regarding the usefulness of FFPs between distinctive categories of workers (e.g., operational *versus* managerial levels); (2) how distinct forms of employment contracts (e.g., temporary *versus* permanent) may play a part in the adoption and use of

FFPs; (3) and, finally, a comparison between organizations in the public and in the private sectors, which might reveal important dissimilarities in terms of family-friendly workplace cultures.

References

1. Albrecht, G. (2003). How friendly are family friendly polices? *Business Ethics Quarterly, 13,* 177–192.
2. Allard, K., Haas, L., & Hwang, C. (2011). Family-supportive organizational culture and fathers' experiences of work–family conflict in Sweden. *Gender, Work and Organization, 18* (2), 141–157.
3. Allen, T. (2001). Family-supportive work environments: The role of organizational perceptions. *Journal of Vocational Behavior, 58,* 414–435.
4. Amah, O. (2010). Family-work conflict and the availability of work-family friendly policy relationship in married employees: The moderating role of work centrality and career consequence. *Research and Practice in Human Resource Management, 18,* 35–46.
5. Ashforth, B. E., Kreiner, G. E., & Fugate, M. (2000). All in a day's work: Boundaries and micro role transitions. *Academy of Management Review, 25*(3), 472–491.
6. Bagilhole, B. (2006). Family-friendly policies and equal opportunities: A contradiction in terms. *British Journal of Guidance & Counseling, 34*(3), 327–343.
7. Bardoel, E. A. (2003). The provision of formal and informal work-family practices: The relative importance of institutional and resource dependent explanations versus managerial explanations. *Women in Management Review, 18*(1/2), 7–19.
8. Beauregard, A. (2007). Family influences on the career life cycle. In M. F. Özbilgin & A. Malach-Pines (Eds.), *Career choice in management and entrepreneurship: A research companion* (pp. 101–126). Cheltenham: Edward Elgar Publishing.
9. Bond, J. T., Galinsky, E., Kim, S., & Brownfield, E. (2005). *2005 National study of employers: Highlights of findings.* Families and Work Institute. http://familiesandwork.org/site/research/reports. Accessed July 05 2015.
10. Brought, P., & O' Driscoll, M. (2010). Organizational interventions for balancing work and home demands: An overview. *Work & Stress, 24*(3), 280–297.
11. Coronel, J., Moreno, E., & Carrasco, M. (2010). Work–family conflicts and the organizational work culture as barriers to women educational managers. *Gender, Work and Organization, 17* (2), 219–239.
12. Daverth, G., Hyde, P., & Cassell, C. (2016). Uptake of organizational work-life balance opportunities. The context of support. *International Journal of Human Resource Management, 27*(15), 1710–1729.
13. de Sivatte, I., & Guadamillas, F. (2013). Antecedents and outcomes of implementing flexibility policies in organizations. *International Journal of Human Resource Management, 24*(7), 1327–1345.
14. den Dulk, L., & de Ruitjer, J. (2008). Managing work-life policies: Disruption versus dependency arguments. Explaining managerial attitudes towards employee utilization of work-life policies. *International Journal of Human Resource Management, 19*(7), 1222–1236.
15. DiMaggio, P. J., & Powell, W. W. (1983). The iron cage revisited: Institutional isomorphism and collective rationality in organizational fields. *American Sociological Review, 48,* 147–160.
16. Dolcos, S. M. (2006). *Managing life and work demands: The impact of organizational support on work-family conflict in public and private sectors.* North Carolina: North Carolina State University.

17. Eby, L. T., Casper, W. J., Lockwood, A., Bordeaux, C., & Brinley, A. (2005). Work and family research in IO/OB: Content analysis and review of the literature (1980-2002). *Journal of Vocational Behavior, 66,* 124–197.
18. Edwards, J. R., & Rothbard, N. P. (2000). Mechanisms linking work and family: Clarifying the relationship between work and family constructs. *Academy of Management Review, 25*(1), 178–199.
19. Eriksson, P., & Kovalainen, A. (2008). *Qualitative methods in business research.* London: Sage.
20. Frye, N., & Breaugh, J. (2004). Family-friendly policies, supervisor support, work-family conflict, family-work conflict and satisfaction: A test of a conceptual model. *Journal of Business and Psychology, 19*(2), 197–220.
21. Galinsky, E., & Stein, P. (1990). The impact of human resource policies on employees. *Journal of Family Issues, 11*(4), 368–383.
22. Galinsky, E., Sakai, K., & Wigon, T. (2011). Workplace flexibility: From research to action. *The Future of Children, 21*(2), 141–162.
23. Giannikis, S. K., & Mihail, D. M. (2011). Flexible work arrangements in Greece: A study of employee perceptions. *International Journal of Human Resource Management, 22*(2), 417–432.
24. Gibbs, G. (2007). *Analyzing qualitative data*. London: Sage.
25. Glass, J., & Fujimoto, T. (1995). Employer characteristics and the provision of family responsive policies. *Work and Occupations, 22,* 380–411.
26. Glass, J., & Estes, S. (1997). The family responsive workplace. *Annual Review of Sociology, 23*(289), 313.
27. Greenhaus, J., & Beutell, N. (1985). Sources of conflict between work and family roles. *The Academy of Management Review, 10*(1), 76–88.
28. Gregory, A., & Milner, S. (2009). Editorial: work-life balance: A matter of choice? *Gender, Work and Organization, 16*(1), 1–13.
29. Grover, S., & Crooker, K. (1995). Who appreciates family-responsive human resource policies: The impact of family-friendly policies. *Personnel Psychology, 48*(2), 271–288.
30. Hopkins, K. (2005). Supervisor support and work-life integration: A social identity perspective. In E. E. Kossek & S. J. Lambert (Eds.), *Work and life integration: Organizational, cultural, and individual perspectives* (pp. 445–468). New Jersey: Lawrence Erlbaum Associates.
31. Kirby, E. L., & Krone, K. J. (2002). The policy exists but you can't really use it: Communication and structuration of work-family policies. *Journal of Applied Communication Research, 30,* 50–77.
32. Kossek, E. E., Lautsch, B. A., & Eaton, S. C. (2005). Flexibility enactment theory: Implications of flexibility type, control, and boundary management for work–family effectiveness. In E. E. Kossek & S. J. Lambert (Eds.), *Work and life integration: Organizational, cultural, and individual perspectives* (pp. 243–262). New Jersey: Lawrence Erlbaum Associates.
33. Kotey, B., & Sharma, B. (2015). Predictors of flexible working arrangement provision in small and medium enterprises (SMEs). *International Journal of Human Resource Management*. doi:10.1080/09585192.2015.1102160.
34. Lewis, S. (2001). Restructuring workplace cultures: The ultimate work-family challenge? *Women in Management Review, 16*(1), 21–29.
35. McCarthy, A., Darcy, C., & Grady, G. (2010). Work- life balance policy and practice: Understanding line manager attitudes and behaviors. *Human Resource Management Review, 20,* 158–167.
36. McCarthy, A., Cleveland, J., Hunter, S., Darcy, C., & Grady, G. (2013). Employee work-life balance outcomes in Ireland: A multilevel investigation of supervisory support and perceived organizational support. *International Journal of Human Resource Management, 24*(6), 1257–1276.

37. McCrea, R., Boreham, P., & Ferguson, M. (2011). Reducing work-to-life interference in the public service: The attributes of participative management as mediated by other work attributes. *Journal of Sociology, 47,* 313–332.
38. McGowan, R. (2009). Managerial discourses of work and eldercare: (Re)producing, resisting, and negotiating boundaries between private and public. *Culture and Organization, 15*(3/4), 307–329.
39. McNamara, T. K., Pitt-Catsouphes, M., Matz-Costa, C., Brown, M., & Valcour, M. (2013). Across the continuum of satisfaction with work–family balance: Work hours, flexibility-fit, and work–family culture. *Social Science Research, 42,* 283–298.
40. Mesmer-Magnus, J., & Glew, D. (2012). Workplace predictors of family-facilitative coworker support. *Journal of Workplace Behavioral Health, 27*(4), 289–310.
41. Monteiro, R., & Domingos, L. (2013). O sentido do direito à conciliação: Vida profissional, familiar e pessoal numa autarquia. *Sociologia, Problemas e Práticas, 73,* 59–77.
42. Osterman, P. (1995). Work/family programs and the employment relationship. *Administrative Science Quarterly, 40,* 681–700.
43. Pasamar, S., & Valle, R. (2015). Antecedents of work–life involvement in work–life issues: Institutional pressures, efficiency gains or both? *International Journal of Human Resource Management, 26*(8), 1130–1151.
44. Pocock, B., Skinner, N., & Williams, P. (2008). Work- life outcomes in Australia: Concepts, outcomes and policy. In C. Warhurst, D. R. Eikhof, & A. Haunschild (Eds.), *Work less, live more? Critical analysis of the work-life boundary* (pp. 44–61). New York: Palgrave Macmillan.
45. Poelmans, S., Chinchilla, N., & Cardona, P. (2003). The adoption of family-friendly HRM policies: Competing for scarce resource in the labour market. *International Journal of Manpower, 24*(2), 128–147.
46. Ryan, A. M., & Kossek, E. E. (2008). Work-life policy implementation, breaking down or creating barriers to inclusiveness. *Human Resource Management, 47*(2), 295–310.
47. Santos, G. G. (2015). Narratives about work and family life among Portuguese academics. *Gender, Work and Organization, 22*(1), 1–15.
48. Santos, G. G., & Cabral-Cardoso, C. (2008). Work-family culture in academia: A gendered view of work-family conflict and coping strategies. *Gender in Management: An International Journal, 23*(6), 442–457.
49. Seeleib-Kaiser, M., & Fleckenstein, T. (2009). The political economy of occupational family policies: Comparing workplaces in Britain and Germany. *British Journal of Industrial Relations, 47*(4), 741–764.
50. Straub, C. (2012). Antecedents and organizational consequences of family supportive supervisor behavior: A multilevel conceptual framework for research. *Human Resource Management Review, 22,* 15–26.
51. Sutton, K. L., & Noe, R. A. (2005). Family-friendly programs and work-life integration: More myth than magic? In E. E. Kossek & S. J. Lambert (Eds.), *Work and life integration: Organizational, cultural, and individual perspectives* (pp. 151–170). New Jersey: Lawrence Erlbaum Associates.
52. Sumer, S., Smithson, J., Guerreiro, M., & Granlund, L. (2008). Becoming working mothers: Reconciling work and family at three particular workplaces in Norway, the UK and Portugal. *Community, Work & Family, 11*(4), 365–384.
53. Thompson, C. A., & Prottas, D. J. (2005). Relationships among organizational family support, job autonomy, perceived control, and employee well-being. *Journal of Occupational Health Psychology, 10*(4), 100–118.
54. Thompson, C. A., Beauvais, L. L., & Lyness, K. S. (1999). When work–family benefits are not enough: The influence of work–family culture on benefit utilization, organizational attachment, and work–family conflict. *Journal of Vocational Behavior, 54,* 392–415.
55. Thornthwaite, L., & Sheldon, P. (2004). Employee self-rostering for work-family balance: Leading examples in Austria. *Employee Relations, 26*(3), 238–254.

56. Townsend, K., McDonald, P., & Cathcart, A. (2016). Managing flexible work arrangements in small non-for-profit firms: The influence of organizational size, financial constraints and workforce characteristics. *International Journal of Human Resource Management*. doi:10.1080/09585192.2015.1136671.
57. Vianen, A., & Fischer, A. (2002). Illuminating the glass ceiling: The role of organizational culture preferences. *Journal of Occupational and Organizational Psychology, 75,* 315–337.
58. Webber, M., Sarris, A., & Bessel, M. (2010). Organizational culture and the use of work-life balance initiatives: Influence on work attitudes and work-life conflict. *The Australian and New Zealand Journal of Organizational Psychology, 3,* 54–65.
59. Wise, S., & Bond, S. (2003). Work-life policy: Does it do exactly what it says on the tin? *Women in Management Review, 18,* 20–31.
60. Yin, R. K. (1994). *Case Study Research: Design and Methods*. Thousand Oaks: Sage.
61. Zartaloudis, S. (2014). The impact of the fiscal crisis on Greek and Portuguese welfare states: Retrenchment before the catch-up? *Social Policy & Administration, 48*(4), 430–449.

Workforce Diversity in Small- and Medium-Sized Enterprises: Is Social Identification Stronger Than the Business Case Argument?

David Starr-Glass

Abstract Historically, workplace diversity has not been considered an intuitive practice. At different times it has been viewed as a moral issue, a legally mandated requirement, or as a combination of both. Today, workplace diversity is usually presented in terms of a rational and pragmatic economic advantage: the so-called business case argument. However, the implementation of diversity always takes place within a social and cultural context and raises question of self-identification and social categorization for those who are in, or who are coming into, the organization. Because of the numbers of employees involved, questions of identity and the impact of diversity are perhaps most acutely encountered in small- and medium-sized enterprises. This chapter considers the business case argument, the paradoxes associated with diversity, and the potential challenges that result when increasing shifts toward diversity are pursued in micro- and small-sized business ventures.

Introduction

A recent survey by the Society for Human Resource Management [SHRM] of 546 senior executives, most of them at the CEO and CFO level, indicated that more than half (55%) had organizational polices that promoted diversity and inclusion either 'strongly' or 'very strongly.' This high level of positive endorsement was consistent for North America (59%) and the Asia-Pacific and Europe region (59%). The same report identified five main rationales for organizations pursuing diversity policies, with most respondents indicating more than one: (a) gaining internal efficacy by

D. Starr-Glass (✉)
University of New York in Prague, Prague, Czech Republic
e-mail: David.Starr-Glass@esc.edu

D. Starr-Glass
Empire State College, State University of New York, Saratoga Springs, NY, USA

D. Starr-Glass
P.O. Box 1154, 91000 Jerusalem, Israel

C. Machado and J.P. Davim (eds.), *Managing Organizational Diversity*,
DOI 10.1007/978-3-319-54925-5_5

tapping into a broader range of backgrounds and skill sets (53%); (b) doing the right thing—that is a matter of fairness and morality (47%); (c) understanding customer diversity and thereby increase sales (43%); (d) being mindful of the diversity of customers and suppliers (42%); and (e) complying with equal opportunity laws and avoiding claims for discrimination (38%) [1].

These figures reflect some of the organizational rationales for actively pursuing workplace diversity, and most of those rationales recognize that diversity leads to a richer organizational inventory of skills, talents, and experiences, all of which can result in economic and marketplace advantages. However, this survey only considered large-scale enterprises. A similar SHRM survey taken the year before also included small- and medium-sized enterprises (SMEs) and the results were significantly different. The survey was completed by more than 1400 senior human resource (HR) professionals and diversity practitioners in North America. It revealed that on average only 30% of all organizations associated with the survey had an official definition of diversity. But the total average is skewed and misleading: large-scale firms are much more likely to have such a definition (43%) than either medium-sized (16%), or small-sized (14%) enterprises [2]. Clearly, although large-scale firms have enthusiastically embraced diversity efforts, that enthusiasm is lacking in SMEs.

It is difficult to get a clear or accurate picture of workforce diversity in the SMEs. Yet, there are fragmentary images in the literature and the composite picture they provide is far from encouraging. For instance, surveying a sample of SMEs in the UK, researchers found 'a distinct lack of awareness or regard for good practice and the risks of potential litigation' and that many firms 'were actively perpetuating direct and indirect discrimination' [3, p. 12]. Five years later, some of the researchers conducted a follow-up study on gender-based diversity in SMEs that revealed a greater degree of implicit organizational sympathy for diversity, a growing management appreciation that diversity could offer positive business advantages, but which nevertheless painted a bleak picture of advances in diversity and inclusion in the workforce [4]. The picture of the SME workforces is hazy, but a lack of diversity is clear among those who created small businesses. In 2010, only 8% of all SMEs in the UK were led by a member of a minority ethnic group and of these almost all (88%) were micro-enterprises [5]. It is also estimated that only about 20% of the UK's 4.8 million SMEs are led by women [6, 7].

The Great Recession of 2008 brought about a profound and protracted downturn in the fortunes of most firms, whether in North America or in Europe. SMEs were particularly vulnerable, partly because of their limited organizational resources and partly because of their difficulties in accessing credit and the financial markets. However, SMEs are also particularly flexible and resilient, and many considered that they would ultimately weather the recessionary storm. In particular, many suggested that the informality and flexibility of their HR management systems allow them to respond more rapidly and decisively than their large-scale counterparts [8–10].

This chapter revolves around a simple but perplexing question. *If* SMEs are particularly susceptible to economic and marketplace downturns, and *if* SMEs do possess HR structures that are flexible and responsive, why in these difficult times have they not embraced diversity efforts more enthusiastically for reasons of pragmatic advantage if not of moral justice or legal compulsion?

This theoretical chapter seeks to answer this question, albeit partially and not completely. A complete answer can only come when research has more clearly illuminated many of the issues that are presently obscure. However, to provide that partial answer the chapter is structured as follows. First, it provides a background sketch of SMEs, outlining their prevalence and significant economic. Second, it explores the sources and consequences of diversity paradoxes and the sets out what is generally known as *the business case* argument for diversity. Third, it examines the evolving ideologies that have shaped our thinking about workforce diversity and the forces that have advanced it. Fourth, the chapter considers the social, psychological, and group dynamics within SMEs that define self-identity and provide a sense of inclusion and membership. Fifth, it discusses the social and group dynamics that limit diversification of existing SME workforces. Finally, it reviews some of the salient considerations, constraints, and convictions associated with SMEs and their diversity efforts.

SMEs: Definitions, Prevalence, and Size Constraints

Obviously, SMEs are classified by their relative size. However, there is considerable variation in the criteria that have been suggested for measuring 'size' and whether absolute or relative values should be employed in differentiating 'smaller' from large-scale operations [11]. A simple headcount approach fails to adequately define small-sized enterprises and tends to obscure their qualitative differences with large-scale corporations.

Many current definitions of a SME consider employee headcount together with the firm's economic activity (say, in terms of annual turnover) and/or its relative financial strength (say, its balance sheet assets). However, economic and financial parameters often prove difficult to determine and are frequently reported inaccurately by SMEs themselves. Nevertheless, a firm's headcount together with its economic and financial information does provide a more realistic, robust, and informative definition of small and medium-sized enterprises. These parameters are reflected in current EU definitions [12–14]:

- **Micro-sized enterprises** have less than 10 workers, an annual turnover of less than €2 million, and balance sheet assets of less than €2 million;
- **Small-sized enterprises** employ between 10 and 49 people, have an annual turnover in the €2–10 million range, and balance sheet assets in the €2–10 million range;
- **Medium-sized enterprises** employ between 50 and 250 people, have an annual turnover in the €10–50 million range, and balance sheet assets of €10–43 million.

In addition to these quantitative parameters, SMEs are characterized as being legally and economically autonomous—that is of being capable of making their own operational decisions and formulating their own strategies.

In the US, the Small Business Administration has established two widely used size standards: (a) in the manufacturing and mining sectors, small businesses are generally those with less than 500 employees; and (b) in the nonmanufacturing sectors, small businesses are usually those with average annual receipts of less than $7.5 million. However, the determination of small business status—and accordingly the federal support and assistance that might be available—is often determined by specific government policy for different local and national economic sectors [15].

SMEs are by far the most common business entity in both developed and developing economies, and in many economies SMEs contribute the largest amount to the GDP and are the largest source of employment. In Europe, SMEs are invariably referred to as the *backbone of the economy*, and for good reasons. In the EU-28 countries, during the year 2014, there were more than 22.3 million SMEs operating in the nonfinancial sector, which is 99.8% of all economically active businesses in that sector, with micro-, small-, and medium-sized enterprises constituting 93, 6, and 1% of the total respectively [16].

In 2014 micro-, small-, and medium-sized enterprises contributed €3.7 trillion in gross value added (GVA) to the EU-28 economy (21, 18, and 19% of total GVA, respectively), with large-scale businesses adding only €2.7 trillion (42%). In the same year micro-, small-, and medium-sized enterprises employed almost 90 million members of the EU-28 labor force (29, 21, and 17% of total employment, respectively), while large-scale businesses employed 44 million people, or slightly less than half of those in SMEs and a third of the total workforce [16].

In the US, the latest business census of 2008, which coincided with the onset of the Great Recession, indicated that small businesses (those with less than 500 employees) produced 46% of the private nonfarm GDP [17]. Employment statistics for 2012 indicate that just over 56 million people (about 49% of the total US workforce) were employed in small businesses. The small business employment figure can be further decomposed into: (a) those working in firms with fewer than 20 employees, just over 20 million (about 18% of the total US workforce); (b) those working in firms with between 20–99 employees, just under 20 million (17%); and (c) those working in firms with 99–499 employees, just over 16 million (14%) [18].

The dominance of SMEs in the labor markets and as a source of employment means that they are central in any effort to increase diversity in the workplace. Certainly, they are recognized as such by policy makers in the EU, who have worked consistently to bring workplace diversity to the forefront and to promote greater inclusion in SMEs. Indeed, just as SMEs are a reliable indicator of national economic health, so too diversity within SMEs is an equally reliable indication of diversity at the national level [19]. EU diversity projects place great emphasis on the *business case* argument for diversity in the workplace, which basically says that increased diversity provides the firm with a range of economic benefits—such as increased talent, knowledge, and experience—all of which lead to a competitive advantage in the marketplace. The business case argument is usually presented as quite unproblematic and is considered crucial in raising the diversity awareness of SME management for 'whom diversity is not necessarily a priority or a spontaneous concern' [20, p. 12].

The Paradoxes of Diversity and the Business Case

At first sight, it would seem that workforce diversity is a rather straightforward concept that can be objectively presented and effectively monitored. However, over its evolution 'diversity'—and for that matter 'inclusion'—has come to mean different things, to different people, at different times. It might, therefore, be helpful to offer a definition of diversity and inclusion, and then to explore the ways in which ambiguity and complexity have set in.

Diversity and Inclusion

Often the terms diversity and inclusion are used synonymously. However, each term reflects a different understanding of what it means to be a member of, and included in, the organization.

Diversity and the Dimensions of Diversity

Diversity is a statistically determined characteristic of a given population that measures the degree to which different classes are represented in it. Of course, since every person is different and unique it might be tempting to infer that any nonclonal population is by its very existence diverse. The difficulty is that distinct classes must be determined before diversity can be measured and this results in two problems. The first is that the class assignment—that is the elements of variation that will constitute the dimensions of diversity—is essentially subjective, always arbitrary, but nevertheless considered important by those who will be determining diversity. The second is that the classes assigned might not correspond to the self-identifications and self-categorizations of those being classified, or seem important to them, leading to their different perceptions or experiences of diversity.

Both of these problems can be reduced by exploring the elements of variation that people feel are important in the way they see themselves and others. Research suggests that the *primary dimensions* of diversity—those that people believe are most important in defining self and others—include gender, ethnicity, race, and to a lesser extent sexual orientation, socio-economic status, age, and mental or physical characteristics. The *secondary dimensions* of diversity are more variable in nature and are believed by most individuals to exert a less overt, or significant, influence on matters of self-identity self-classification, and on the identification and classification of others. As might be expected, secondary dimensions include a wide range of characteristics including (but not limited to) educational background, religion, first language, and family status [21–26].

In determining organizational diversity, various combinations of primary and secondary dimensions could be used. However, because of the historical pathways that diversity has taken, the diversity dimensions actually used are (normally) the same as those prescribed by existing legal requirements. Legally defined diversity dimensions are those that characterize demographic classes that have been afforded legislated protection against discrimination and unfavorable treatment. In the EU, this currently includes gender, sexual orientation, disability, age, religion or beliefs, and racial or ethnic origin. Workplace *diversity* can then be measured as the distribution of differences among workplace participants with respect to these defined classes. This provides an index of compositional variety within the group.

Although measuring the extent of diversity is useful, the organization must still determine against what the index should be compared. Should it be compared to other organizations, economic sector demographics, or the demographics of the ambient community? Should it be compared to organizational goals such as increasing or maximizing diversity ? The index also is a statistical representation of diversity and does not provide any insight into the *consequences* of diversification—changes in levels of attitudinal *separation* between group members, or of perceived power and influence *disparity* between them. Organizations might not consider these questions and simply use their measures of diversity in a defensive manner to infer that their diversity statistics demonstrate the absence of institutional discrimination.

Inclusion and Inclusiveness

Inclusion is an internal perception of those within the organization; whereas, diversity tends to be a managerial response to externally imposed conditions. Inclusion is the extent to which organizational participants consider themselves to be involved and *included* within the organization. From this perspective, inclusion is primarily about perceived social identification with, and membership of, the organization in which they work. Inclusion is an internalized sense of employee belonging to the workplace that is stronger than the sense of identification with external nonorganizational groups and groupings—at least when the individual is presently situated in the organization, or recognizes organizational membership to be salient.

Although a sense of inclusion is developed by the organizational participant, it can be prompted and promoted by the organization. Thus, *inclusive HR practices* try to create organizational and workplace environments that are characterized by a sense of community, mutual respect, and collective responsibility. These HR practices often center on furthering the integration of those with disparate cultural and ethnic heritages into the fabric of the organization, and of promoting individual difference as a source of shared insight and opportunity within the organizational community [27–31].

The Diversity Paradox

It has been pointed out that *diversity*—just like discrimination, multiculturalism, and inclusion—is a rather nebulous construct that is 'fuzzy in meaning, and vary depending on the geographical, cultural, political, and organizational context' [32, p. 245]. For some, the diversity construct might be fuzzy and fluid in nature. For others, the appeal and advantages of diversity in the workplace might be self-evident.

However, the operationalization of the diversity, the development of policies designed to promote diversity, and the diversity management enactments of HR departments are all challenging. In part, these challenges rest on what appears to be a set of interlocking paradox about the nature and the consequences of diversity. These paradoxes arise because all organizational efforts connected with diversity are rooted in a matrix of social, psychological, and ideological perspectives that are often implicit and not fully articulated, but which nevertheless lead to paradoxical, unintended, and counter-intuitive outcomes [33].

Scholars have framed these paradoxes in slightly different ways, but each is grounded in the recognition that diversity ideologies are often not congruent with the behavioral outcomes and social realities that follow diversity implementation.

- **Diversity and Like-Minded Others** Hackett and Hogg [34] highlight the paradoxical dynamics of diversity efforts in organizations by noting that 'those who value diversity as a group attribute may strive to be with and live in communities of like-minded others, people who also value diversity' (p. 416). Thus, as diversity efforts are increased those who favor diversity tend to become affiliated with the group, while those who are not diversity-minded are self-excluded. The resulting group may be very predisposed to the notion of diversity, but its very sameness precludes or significantly reduces its own diversity. Like-minded communities are usually not very diverse.
- **Diversity and Clonal Replication** In a similar manner, Herring and Henderson [35] see a tension and a potential paradox between: (i) a fundamental and genuine understanding held by the organization that 'without linking it to concerns about access, equity, parity, and opportunity, the usage of the term diversity is hollow' (p. 22), and (ii) the human psychological and sociological reality that 'most organizations attempt to reproduce themselves by selecting people who are similar to the members or constituencies they already have' (p. 23). Here the organization, in honestly desiring to achieve expansive diversity and equity in an abstract sense, acts (through its recruitment, selection,

and reward systems) in way that preserves the status quo. By definition, clonal replication cannot produce diversity.

- **Diversity and Divergent Behavior** DiTomaso, Post, and Parks-Yancy [36] argue that the somewhat abstract acceptance of identifiable difference and of incorporating them into organizational settings is inevitably leads to an increase in differing human needs, behaviors, and goals. They note that 'the critical question is [i] whether it is possible to retain a sense of difference without [ii] difference taking on a negative connotation… in other words, is pluralism possible without an unhealthy competition for resources?' (p. 492). Diversity is often seen in terms of adding varietal richness into organizational *blend*; however, diversity can also give rise to a growing sense of disparity and separation for those in the existing in-group. Diversity can be viewed in a descriptive compositional sense, but diversity also has behavioral consequences and some of those can be negative.
- **Diversity and Discontent** Milliken and Martins [37] in their attempt to 'find common threads' in the research literature associated with group composition and outcomes—specifically whether greater diversity leads to higher productivity—encountered a problem with the *business case* argument for diversity. This generally claims that all diversity is associated with positive and advantageous outcomes. Reviewing recent management research, however, they concluded that diversity 'appears to be a double-edged sword, [i] increasing the opportunity for creativity as well as [ii] the likelihood that group members will be dissatisfied and fail to identify with the group' (p. 403). Although often promoted in a unilateral way, diversity has a double-sided nature. In some contexts and at some levels, diversity is *seen as variety* (with a positive connotation). In other contexts and at other levels, diversity is *seen as separation* (with a negative connotation).

These paradoxes come about because diversity is not the pre-existing norm in social groups and business organizations. Social groups tend to preserve their identities and perpetuate themselves because they are homogenous. Self-categorization theory suggests that existing groups engender group cohesiveness by encouraging members to see themselves in terms in-group membership, creating a dynamic in which there is a 'tendency to favor members of one's own group over members of other groups' [38, p. 10]. Moves to upset this stability through the increased diversity of group members—especially the enforced diversification of the group—are perceived as threats to in-group homogeneity, cohesiveness, stability and the benefits that individual members have previously accrued through membership, self-identification, and social status.

The Business Case for Diversity

Nowhere are the paradoxes of workplace diversity seen more clearly than in the SME. By definition, these firms are *small*—that is, they employ small workforces that place limits on the extent to which diversity can be expressed, even if desired. SMEs usually lack an extensive HR function that can advise on diversity initiatives, create infrastructures for articulating diversity objectives, or implement effective diversity programs. More significantly the *smallness* of the SME in terms of its number of organizational participants creates in-group situations in which existing group members seek to preserve their internal cohesion and to perpetuate their assumed identity through selective—and often unconscious—recruitment, selection, and retention practices that preserve the status quo.

These dynamics take place at the level of the individual and the group. The SME transcends both of these levels; it is a separate entity—not simply a legal entity, but a distinct organizational entity with its own set of dynamics and goals. Diversity presents challenges and opportunities for individual and group members; however, are these similar to the challenges and opportunities of SMEs, or does diversity always produce advantageous outcomes for them?

From the perspective of *diversity as variety*, it is clear that firms who encourage workplace diversity can derive a number of advantages: (a) they can access a greater range of talent in the labor markets; (b) they can come to a better understanding of their customers, suppliers, and competitors in an increasingly multicultural environment; and (c) they can arrive to more relevant and informed decisions in an more complex, diverse, and globalized world. Even in 1990, there were those who appreciated that, at a national level (the US), for 'a country seeking competitive advantage in a global economy, the goal of managing diversity is to develop our capacity to accept, incorporate, and empower the diverse human talents of the most diverse nation on earth'[39, p. 117].

Historically, the *business case* proposition for diversity developed as a positive alternative to state-legislated affirmative action programs, which were predicated on norms of social justice and equality. The business case argument suggested that diversity was a valuable resource that could be used by organizations to enhance their effectiveness in a changing world. A growing acceptance of multiculturalism, the realization that globalization was a power to be reckoned with, and the conviction that these represented economic opportunities shifted diversity arguments away from moral, ethical, and judicial realms to economic ones. Diversity shifted from equality-based perspectives with their 'moral reasoning grounded in deontology' to perspectives 'crafted by utilitarian arguments' [40].

In the EU, the business case has been strongly advocated as the basis and justification for increased diversity [20]. Diversity and its management are understood to be connected with improved economic performance, enhanced social performance, and better alignment with demographic and business—and, of course, diversity management is in compliance with EU law [19]. Interest in the business argument has been intense among business people, management practitioners, and academics; indeed, to stimulate further academic and research interest the UK's Department for Business Innovation and Skills has recently compiled a comprehensive survey of the literature [41].

Although usually presented as quite unproblematic, the business case for diversity does have its critics both in North America and in the EU. Some question the impact that it has actually had on business decision-makers and on HR practice [42]; others believe that organizations should go beyond utilitarian rationales and reframe their diversity consideration in terms of values, virtues, and ethics [40, 43]; while a few have tried to understand the impact the diversity case has made on SMEs [44].

An Unfolding History of Diversity Ideologies

The shifting landscape of workplace diversity mirrors changes in dominant ideological systems. In part, these ideological systems are rooted cognitively, but more importantly they emerge from widely held and commonly shared understandings and perspectives of social identity groups. Nkomo and Hoobler [32] define *diversity ideology* as the 'societal beliefs and attitudes about nondominant racioethnic groups, their status, and how they should be incorporated into the society or nation' (p. 246). They use this definition to analyze the dominant racial and ethnic ideologies in the US throughout the twentieth and early twenty-first centuries, connecting these ideologies to the prevailing attitudes of employers and the practices of their HR departments. Their analysis suggest four eras in an evolving understanding of workplace diversity.

The First Era: Active Exclusion and Discrimination

This period extended well into the second half of the twentieth century and mirrored the broad socially endorsed attitudes toward race, ethnicity, and gender. Paradoxically, the US from its inception was a country of immigrants; however, increased immigration and the inflow of different ethnic identity groups in the early 1900s was seen as a threat to the perceived supremacy of the dominant majority, which self-identified as White, Protestant, and male. Dominance and supremacy was not understood in numerical terms, but as a preordained reality based on an uncontested superiority of *race* which was rigidly preserved through social, cultural, and legal mechanisms. These attitudes and assumptions informed HR practices that were overtly discriminatory and segregationist.

More *pragmatically* informed HR practices sometime came into play when there was a pressing need for labor. Black employees were actively incorporated into the labor force and vocational training and educational programs were created for them. During this period, Black HR managers were often hired because they were presumed to be more sensitive to the issues of their co-workers. Interestingly, Delton [45] makes a connection between the growth of the Human Relations School of Management and rise of nondiscriminatory hiring practices after World War II, and notes that 'recognizing the role corporate actors played in integration offers an important corrective to the current historical literature on workplace integration, which is based largely on post-1964 sources and which emphasizes grassroots activism' (p. 295).

The Second Era: Legally Mandated Equal Opportunity

Shifting political and social attitudes toward workplace and institutional discrimination brought about the enactment of the federal Civil Rights Act of 1964. Although in some sectors of society there had been a growing appreciation of equal consideration from legal and moral perspectives, these did not usually translate into equality in social arrangements, educational opportunity, or employment. The Civil Rights Act of 1964 shifted things from a matter of conscience to a matter of legal compliance. Specifically, Title VII of the Act made it illegal for employers with more than 15 employees to discriminate in hiring, terminating, promoting, compensating, job training or any other term, condition or privilege of employment on the basis of sex, age, race, color, religion, and national origin [46, 47].

Despite the differing constructions that employers placed on Title VII—and their hesitancy and reluctance to follow the spirit, if not the letter, of the law—the passage of the Equal Employment Opportunity (EEO) Act of 1972 clarified many issues and strengthened the resolve of the federal government to ensure equal opportunities in employment. The Act provided a mechanism whereby lawsuits could be initiated, with federal government support, by those who had encountered discriminatory employment practices in the hiring process or in the workplace. This legislation placed a very real economic cost of noncompliance against the organization's philosophical and rhetorical considerations of equal employment. The law pushed employers to establish EEO practices at all levels of their HR process, and encouraged them to formally adopt what many have called *color blind* approaches to the recruitment, selection, training, promotion, and retention of their workforce [48, 49].

The Second Era of diversity reflected a general social revision of diversity and discriminatory ideologies, albeit a revision that was catalyzed (if not actively pushed and forced) by legal enactments. Increasingly—although halting and not universally subscribed to—there seemed to have been a movement away from the norms of ingrained supremacy and discrimination to a more nuanced understanding in which issues of fairness and equity were more clearly articulated and preferred. The preference for greater equity and inclusion were clearly expressed in the areas of employment and the workplace, which might have been seen as relatively neutral and dispassionate arenas of economic activity, rather than as more contentious arenas of social engagement [50, 51].

The Third Era: Multiculturalism and Diversity Management

Diversity ideologies had generally included a *melting pot* metaphor, in which diversity was gradually diffused and incorporated into society, providing it with a greater cultural and ethnic richness [52]. In contrast, *multinationalism* has the root metaphor of a *tossed salad*, where different cultural identities are preserved intact and remain discernable although they contribute to (in a metaphoric sense) the color, texture, and vitality of the whole.

The term *multiculturalism* originated in Canada, where it featured as a guiding principle and driving philosophy in the country's official social and immigration policy as early as 1971. The concept slowly spread to the US in the late 1980s and early 1990s [32, 52]. Multiculturalism places importance on the integrity of different cultural and ethnic identity groups, recognizes their legitimacy, and accepts them as contributing to a pluralistic society. It values the different perspectives and experiences of all social identity groups—whether differentiated culturally, racially, ethnically, or by gender—and sees each as contributing uniquely to the cohesiveness of the whole society, not as being in any way subservient, inferior, or privileged.

Multicultural approaches increasingly led to organizations seeing diversity as an integral aspect of the human resources and human capital to which they had access. It was appreciated that the diversity of these assets had to be managed and *diversity management* became a significant aspect of organizational rhetoric and HR practice toward the end of the 1990s. The conceptual shift moved diversity away from an externally orientated legal compliance to internally focused issue of effectively managing diversity [39, 53].

The Fourth Era: Post-racial Understanding and Inclusiveness

Multiculturalism recognizes the experiences and perspectives of different identity groups and seeks to balance these into a holistic understanding of society, or the workforce. However, many critics sense that by adhering to 'difference' the multicultural enterprise begins with a premise that has ultimately caused its downfall. They point to a society that is still fractured along socially identifies group lines, with a disproportional accumulation and retention of power and privilege. They point to the possibility of moving beyond the multicultural boundaries that have designated 'different but equal' to a new post-multiculturalism that accentuates equality and eliminates difference; that accentuates the individual and not his/her historical group origins or putative social affiliations.

The post-racial argument for inclusion holds that, by and large, historical groupings—such as race and ethnicity—have waned, or have become essentially irrelevant. However, such a position seems strangely at variance with the reality that many in society and in the workforce experience, or believe that they encounter. Nkomo and Hoobler [32], considering the prospect of a post-racial society, question 'whether this can ever happen, or if this is simply a privileged or utopian view of the world and institutions' (p. 253). Others, such as Fryberg and Stephens [54], argue for a more expanded and critical 'science of diversity' in order to 'brings arguments about ideologies and their effects out of the realm of "sincere ignorance" and "conscious stupidity" and into the realm of research-based observations and solutions' (p. 118).

Many, if not most, question the extent to which a post-multiculturalist and post-racial ideology accurately describe the state of affairs in our contemporary society, the sentiments of those in the workforce, or the understandings of business organization. It seems to most observers that we have actually not entered the Fourth Era. However, it has been rightly pointed out that one impact of Fourth Era thinking has been on the rhetoric, not necessarily on the practices, of HR departments which has increasingly emphasized the need for *inclusion*.

The trajectory taken by diversity ideologiesis clearly shaped by a unique set of social, cultural, and political forces. It would be incorrect to assume that it must pass through the four eras indicated above. However, this brief survey of the US experience may provide useful insight into the evolution of diversity thinking and diversity management in non-US societies.

Identities—Individual, Social, and Organizational

The number of individuals that the firm employs is obviously a key part of SME classification. However, focusing on employee headcounts should not obscure the reality that the SME is a discrete *social collective*, not simply an assemblage of unconnected individuals. SMEs, like all other social organizations, are confronted with issues of articulating their own distinctive identities and with the processes through which organizational members, and prospective members, align their own understandings of social and personal identity with the identity of the firm.

Social Identity and Self-categorization

There are two different theories of identity: *identity theory*, which is grounded in the structural symbolic interaction approach taken by Stryker [55], and *social identification theory*, which has its origins in Tajfel's [56] recognition of social categorization. Although each has acquired its own set of meanings, some have suggested that both theories have more similarities and overlaps than dissimilarities. Generally, identity theory has tended to define identity in terms of the roles that individuals understand they will assume and the meanings that they attach to those roles; whereas, social identity theory has considered identity in terms of the social categories that individuals identify, the degree to which they see themselves belonging to these categories, and the social meaning that they attach to them [57].

Social identity and self-categorization theories both understand that individuals, as social actors in a social world, can neither define a unique and personal identity, nor construct a comprehensive narrative of self, in isolation. The creation of identity is continuously mediated and confirmed by the process of comparison with the other. Social identity and self-categorization theories are essentially similar and both provide ways of understanding how questions such as 'who am *I*?' and 'what is important to *me*?' can be convincingly reframed by the individual, as a social actor, in terms of the more expansive 'who are *we*?' and 'what is important to *us*?' [58, 59].

Individual and social identities work together dynamically, with each setting and then confirming goals that are considered to be contextually salient. As Haslam, van Knippenberg, and Platow [60] note, 'the self can be defined in individual- or group-based terms… [*individual goals*] focus and give meaning to personal identity and serve as a basic for organizing and coordinating (potentially) disparate behaviors… [*group goals*] help to direct and give meaning to shared social identity which is used as a framework for coordinating and organizing the behavior of (potentially) disparate individuals' (p. 51).

Organization affiliation requires an interrogation of both self-identity and self-classification. Affiliation can only be successful—for the individual and for the organization—if the in-comer recasts a personal identity and classification in ways that align with the present organizational in-group. Successful redefinition of personal identity produces a sense of in-group identification and of belonging for the in-comer, and a sense of affirmation and cohesion for the existing in-group members.

A sense of belonging, and of wishing to continue to belong, helps to reduces existential uncertainty and increases the predictability of the future for the members of the group. In-group identification also increases self-esteem of members because their views and behaviors, which *others* have now adopted in order to conform to the group's norms, are now seen as having been affirmed and validated. However, in perpetuating their distinctiveness, identity, and cohesiveness the in-group tends to identify and to define contrasting and problematic out-groups. To maintain clarity of 'who *we* are' and 'what is important to *us*,' we set up an oppositional 'who *they* are' and 'what is important to *them*.'

Importantly, individuals possess multiple social identities and self-categorizations and the saliency of these is contextually determined. Changes in situational factors may lead to new questions of identity and categorization that were previously unconsidered, or which were not considered either important or relevant in the past. For example, in contexts of increasing diversity within the organization, *identity-based fault-lines* may develop within the group, dividing it into a number of new and relatively homogeneous subgroups that are aligned with demographic characteristics of individuals (color, ethnicity, gender, etc.), which have only now become salient.

For example if eight men work in a micro-enterprise they may all tend to self-categorize as members of the firm. However, if the firm has eight workers four of whom are men and four women, the group may still predominately self-categorize as *firm members*, but there is also a new identity category of *gender* available to them and the creation of a potential fault-line. If the men are all old, White, and Christian and the women are all young, Black, and Muslim there is a significant multi-reinforced identity-based fault-line that may well split the group into two. The split could potentially impair collective performance, inhibit collaborative efforts, and bring into question a sense of firm identity and affiliation. In this case, the micro-enterprise can be said to possess 'diversity' (a representation of different genders, races, and religions), but that has not necessarily improved its performance, its success, or its organizational identity. If race and religion had been more evenly divided among the age classes (one White Christian female, one White Christian male, one White Muslim female, etc.) the same number of identity-based fault-lines would be present, but these would run *across* the group rather than all lie on top of one another. This would presumably make the fault-lines less susceptible to fracture.

It is important to note that identity-based fault-lines do not necessarily split the group; however, they do represent lines of potential fracture that might become active in the future. In the example given the fault-lines are accentuated (or reinforced) because *each member* of the subgroups has the same gender, race, and religion. Of equal importance, it should be noted that individuals can *benefit significantly* from the resulting complexities of social identity that occur with group diversity. Group members often appreciate the increased complexity of their self-categorizations and respond by sharing previously unutilized skills and knowledge that contribute positively to the group's performance.

Organizational Identity

The emerging *organizational identity* of the SME is not a matter of managerial rhetoric, but of what its various stakeholders (particularly organizational participants) come to recognize as the central and enduring values that differentiate the firm from other similar enterprises [61–63]. Organizational identity, and the subsequent self-identification of organizational participants with that identity, provides a mechanism for developing social cohesion within the enterprise and a means through which organizational members can answer questions of identity, classification, and affiliation [64, 65].

The construct of organizational identity differs from that of organizational culture; however, both have a reciprocal, dynamic, and synergistic relationship. Organizational culture is the ongoing articulation, elaboration, and refinement of the (succinct) organizational identity narrative, which shapes the behavior and thinking of organizational participants and which is transmitted to those who enter the organization.

Whereas organizational identity may answers questions of 'who are we?' and 'what is important to us?' in the cognitive and affective domains, organizational culture suggests answers for the behavior questions of 'what do we *do*?' and 'why do we do *that*? Although organizational identity and the self-identification of organizational participants are different, they also seem to be reciprocally connected with both necessary pre-requisites for the emergence of other related constructs such as *organizational commitment* and *organizational loyalty* [66, 67].

Considerations, Caveats, and Conviction

There is little disagreement that diversity is a significant and pressing issue in the twenty-first century. When asked, most individuals agree that diversity in the workplace is a reasonable expectation and that discrimination is offensive and runs counter to their sense of fairness. When asked, organizations—large-scale firms and SMEs—claim that they wholeheartedly endorse the notion of diverse and inclusive workforces. And undoubtedly these claims are sincere. For the rational business it would seem that the proposition of a diverse workforce is predominantly a matter of economic sense; a matter of responding to a convincing business case for diversity.

However, this is a particularly modern perspective. Historic attitudes and responses to diversity, and for that matter many contemporary attitudes, are not grounded in an economic calculus. Instead, they embody understandings of social identity and social cohesion that are perhaps more human, behaviorally centered, and absent from the assumed rationality of that pale and spectral being *homo economicus*. Any historical review of diversity and discrimination illustrates the power of social forces that had to be overcome. But, more than that, any historical review offers insights into the present day dynamics of resistance to increasing organizational diversity [68, 69].

SMEs by definition include a small number of participants. In this chapter it is argued that, unlike large-scale firms, the internal dynamics of SMEs are much closer to those of small groups. As a result, increased diversity within the SME might: (a) result in the (problematic) reconsideration of individual identity and self-classification; (b) endanger the existing self-esteem and cohesiveness of the current in-group; and (c) lead to a proliferation of potential identity-based fault-lines. All of these outcomes might be regarded with varying degrees of concern by SME in-group members and by those who regulate group diversity. This concern might lead to the conclusion—explicitly voiced, or subconsciously held—that moves toward increased group diversity are problematic and should be postponed and reconsidered, if not necessarily resisted [70–72].

In this chapter it is speculated that although SMEs are embedded in the social, cultural, and political matrices of their national states their internal dynamics render them somewhat resistant to external social shifts. The progression in SME diversity appreciation might also be slowed by their preoccupation with the sharper issues of existential survival and growth, their wish to avoid the perceived risks in new HR recruitment and selection practices, and general sense that 'diversity is not necessarily a priority or a spontaneous concern' [20, p. 12].

Of course there are a number of limitations and caveats to this rather pessimistic diversity outlook. Many of these have to do with what might be considered yet another paradox. While diversity may be limited *within* SMEs it is expansive *between* SMEs.

- **Firm Size** There are difficulties in defining SMEs by headcounts, but firm size nevertheless plays a crucial part in firm dynamics. The actual number of those in the workforce places limits on the extent to which diversity can be expressed. But more than that, group size is an element in its cohesive and of the number of fault-lines that can develop as diversity increases [72]. Based on these factors, it would seem likely that micro-enterprises will find diversity shifts more problematic than small-sized ones, and that small-sized enterprises will be more challenged than medium-sized firms.
- **Business Sector** SMEs are engaged in multiple business sectors. Some of these sectors—for example those engaged in R&D, technology, innovation, and knowledge sectors—may be more responsive the business case argument, since diversity will increase their talent and innovation pool. Certainly these results have been reported for SMEs in this sector. Increased diversity in boards of directors and top management teams has been associated with higher product and process innovation in Canadian and Spanish SMEs [73, 74]. It is speculated that diversity efforts in these sectors might be more successful than in others where there may be little demonstrable advantage in talent diversity.
- **Founder Personality** It is suggested here that increased diversity in SMEs will to a great extent lie in the *personal* attitudes and dispositions of those who have founded and who are constantly involved in their SMEs, rather than in the distanced and diffused *impersonal* forces and HR policies that can develop in large-scale organization. These *personal* attitudes and dispositions include tolerance for ambiguity, achievement motivation, social identity, and self-categorization. This is not to deny the complexity of behavior within the SME, or to privilege individuals as autonomous and to focus exclusively on the psychological traits of decision-makers [75]. But, as Levison [76] noted for the SME founder the business is essentially an 'extension of himself, a medium for his personal gratification and achievement above all… [if] concerned about what happens to his business after he passes on, that concern usually takes the form of thinking of the kind of monument he will leave behind' (p. 91). Depending on personal attitudes and goals, some SME founders—men and women—may be motivated to leave a living monument in the form of a diverse and inclusive business. They might also be motivated to create an enterprise that reflects and furthers their own social identity.
- **Ownership and Governance Structure** Formal (*contractual*) governance systems are the norm in large-scale enterprises, but these are often replaced by more fluid informal (*relational*) systems in SMEs. Relational governance and subsequent governance behaviors are based on mutual trust, recognizing and respecting *the other*, and patterns of mutual reciprocity [77]. Reviewing her work on Dutch family-based firms, Uhlaner [78] characterizes relational

governance as a vehicle through which 'owners share the same vision of the firm… take a long-term view in their investments and in firm strategy, the degree to which they are committed to the same goals… express an emotional or affective attachment to the firm' (p. 29). It may well be that relational governance patterns strengthen and reinforce in-group attitudes and behavior, particularly about diversity and diversity selection. Equally, it might be that SMEs with contractual governance systems find it easier to arrive at 'balanced and impartial' senior level decisions and advance HR initiative to increase diversity.

In beginning this chapter, it was acknowledged that the relationship between adopting the business case argument for diversity and preserving the conservative impulse of social identification could not be fully resolved. There are still too many unknowns. More research is urgently needed to understand the attitudes that SME participants hold about diversity and about the dynamics that motivate their choices and responses. More research is needed to understand the diversity roles that SME founders, boards of directors, top management teams, and employees see themselves as pursuing. We need to understand their priorities, the trade-offs that they believe they are making, and the reasons for their decisions or nondecisions.

In 2004, the EU established national Diversity Charters through which organizations voluntarily commit to promote diversity and equal opportunities in the workplace, regardless of age, disability, gender, race or ethnic origin, religion or sexual orientation. Ten years later, 13 national Diversity Charters have been established bringing together 7100 European organizations and 13.6 million employees. A recent survey of more than 1100 signatories revealed that they were predominantly (70%) SMEs, and more specifically micro-enterprises (40%) with less than 10 employees [79].

This demonstrates a strong conviction of even the smallest SMEs to voluntarily commit to diversity in their workplaces. Organization size, the social identification of their existing members, and numerous other factors can positively impact a commitment to embrace diversity. These factors, however, can also operate to negate business case arguments for diversity, even although the business case is not without its problems. It would certainly seem that, despite the very real difficulties that SMEs face in advancing diversity, there is a strong and resolute commitment to the concept of diversity, and to making diversity and inclusion an integral part of the SME's internal landscape.

References

1. SHRM. (2009). *Global diversity and inclusion: Perceptions, practices, and attitudes.* A survey report by the Economist Intelligence Unit for the Society for Human Resource Management. Alexandria, VA: Author.
2. SHRM. (2008). *2007 State of workplace diversity management.* A survey report by the Society for Human Resource Management. Alexandria, VA: Author.

3. Woodhams, C., Lupton, B., & Raydon-Rennie, S. (2004). *Gender-based equal opportunities in SMEs: Establishing policy and practice*. Report published by Manchester Metropolitan University Business School.
4. Woodhams, C., & Lupton, B. (2009). Analysing gender-based diversity in SMEs. *Scandinavian Journal of Management, 25*(2), 203–213.
5. IFF Research. (2011), *BIS small business survey 2010: Minority Ethnic Group (MEG) led businesses boost*. Report prepared for the Department of Business, Innovation and Skills (BIS). London, UK. Retrieved from https://www.gov.uk/government/uploads/system/uploads/attachment_data/file/32222/11-1079-bis-small-business-survey-2010-meg-led-businesses-boost.pdf
6. Carter, S., Mwaura, S., Ram, M., Trehan, K., & Jones, T. (2015). Barriers to ethnic minority and women's enterprise: Existing evidence, policy tensions and unsettled questions. *International Small Business Journal, 33*(1), 49–69.
7. Carter, S., Ram, R., Trehan, K., & Jones, T. (2013). *Diversity and SMEs*. Enterprise Research Centre White Paper No. 3. Retrieved from http://www.enterpriseresearch.ac.uk/wp-content/uploads/2013/12/ERC-White-Paper-No_3-Diversity-final.pdf
8. Kitching, J., & Marlow, S. (2013). HR practices and small firm growth: Balancing informality and formality. In G. Saridakis & C. L. Cooper (Eds.), *How can HR drive growth?* (pp. 26–45). Cheltenham, UK: Edward Elgar Publishing.
9. Lai, Y., Saridakis, G., Blackburn, R., & Johnstone, S. (2016). Are the HR responses of small firms different from large firms in times of recession? *Journal of Business Venturing, 31*(1), 113–131.
10. Sheehan, M. (2014). Human resource management and performance: Evidence from small and medium-sized firms. *International Small Business Journal, 32*(5), 545–570.
11. Gibson, T., & van der Vaart, H. J. (2008). *Defining SMEs: A less imperfect way of defining small and medium enterprises in developing countries*. Washington, DG: Brookings Institute. Retrieved from http://www.brookings.edu/~/media/research/files/papers/2008/9/development-gibson/09_development_gibson.pdf
12. Centre for Strategy & Evaluation Services. (2008). *Evaluation of thresholds for micro-entities*. Report prepared for the Evaluation and Impact Assessment Activities of Directorate General for Internal Market and Services. Sevenoaks, UK: Author. Retrieved from http://ec.europa.eu/internal_market/accounting/docs/studies/micro_entity_en.pdf
13. Centre for Strategy and Evaluation Services. (2012). *Evaluation of the SME definition: Final report and executive summary*. Sevenoaks, UK: Author.
14. European Commission. (2005). *The new SME definition: User guide and model declaration*. Enterprise and Industry Publication. Luxemburg: Author.
15. SBA. (2012). *Table of small business size standards matched to North American industry classification system codes*. Washington, DC: Small Business Administration. Retrieved from https://www.sba.gov/sites/default/files/files/Size_Standards_Table.pdf
16. Eurostats. (2015). *Business economy - size class analysis*. European Commission. Retrieved from http://ec.europa.eu/eurostat/statistics-explained/index.php/Business_economy_-_size_class_analysis
17. Kobe, K. (2012). *Small business GDP: Update 2002–2010*. Report for the US Small Business Administration. Retrieved from https://www.sba.gov/content/small-business-gdp-update-2002-2010
18. Caruso, A. (2015), *Statistics of US businesses employment and payroll summary: 2012*. US Department of Commerce, Economics and Statistics Administration. Washington, DC: US Census Bureau. Retrieved from http://www.census.gov/content/dam/Census/library/publications/2015/econ/g12-susb.pdf
19. Hajjar, L. & Hugonet, C. (2015), *Diversity within small and medium-sized enterprises: Best practices and approaches for moving ahead*. Directorate-General for Justice and Consumers. Luxembourg: Publications Office of the European Union. Retrieved from http://bookshop.europa.eu/en/diversity-within-small-and-medium-sized-enterprises-pbDS0215862/

downloads/DS-02-15-862-EN-N/DS0215862ENN_002.pdf?FileName=DS0215862ENN_002.pdf&SKU=DS0215862ENN_PDF&CatalogueNumber=DS-02-15-862-EN-N
20. European Commission. (2015). *Practical guide to launch and implement a Diversity Charter*. Directorate-General for Justice and Consumers, Directorate D—Equality. Brussels, Belgium: European Commission. Retrieved from http://ec.europa.eu/justice/discrimination/files/emerging_charter_guide_en.pdf
21. Baker, D. L., Schmaling, K., Fountain, K. C., Blume, A. W., & Boose, R. (2016). Defining diversity: A mixed-method analysis of terminology in faculty applications. *The Social Science Journal, 53*(1), 60–66.
22. Banks, K. H. (2009). A qualitative investigation of white students' perceptions of diversity. *Journal of Diversity in Higher Education, 2*(3), 149–155.
23. Hon, L. C., Weigold, M., & Chance, S. (1999). The meaning of diversity among the professoriate. *Journalism & Mass Communication Educator, 54*(1), 51–68.
24. Loden, M. (1996). *Implementing diversity*. Boston, MA: McGraw-Hill.
25. Mazur, B. (2010). Cultural diversity in organisational theory and practice. *Journal of Intercultural Management, 2*(2), 5–15.
26. Unzueta, M. M., Knowles, E. D., & Ho, G. C. (2012). Diversity is what you want it to be: How socio-dominance motives affect construals of diversity. *Psychological Science, 23*(3), 303–309.
27. Ely, R. J., & Thomas, D. A. (2001). Cultural diversity at work: The effects of diversity perspectives on work group processes and outcomes. *Administrative Science Quarterly, 46*(2), 229–273.
28. Nishii, L. H. (2013). The benefits of climate for inclusion for gender-diverse groups. *Academy of Management Journal, 56*(6), 1754–1774.
29. Roberson, Q. M. (2006). Disentangling the meanings of diversity and inclusion in organizations. *Group and Organization Management, 31*(2), 212–236.
30. Holvino, E. H., Ferdman, B. M., & Merrill-Sands, D. (2004). Creating and sustaining diversity and inclusion in organizations: Strategies and approaches. In M. S. Stockdale & F. J. Cosby (Eds.), *The psychology and management of workplace diversity* (pp. 245–276). Malden, MA: Blackwell.
31. Shore, L. M., Chung-Herrera, B. G., Dean, M. A., Ehrhart, K. H., Jung, D. I., Randel, A. E., et al. (2009). Diversity in organizations: Where are we now and where are we going? *Human Resource Management Review, 19*(2), 117–133.
32. Nkomo, S., & Hoobler, J. M. (2014). A historical perspective on diversity ideologies in the United States: Reflections on human resource management research and practice. *Human Resource Management Review, 24*(3), 245–257.
33. The B Team. (2015). *The diversity paradox: Capturing the value of difference by looking beyond the numbers*. Report produced in partnership with The Boston Consulting Group. New York, NY: Retrieved from http://issuu.com/the-bteam/docs/the-diversity-paradox-150114?e=15214291/11024239
34. Hackett, J. D., & Hogg, M. A. (2014). The diversity paradox: When people who value diversity surround themselves with like-minded others. *Journal of Applied Social Psychology, 44*(6), 415–422.
35. Herring, C., & Henderson, L. (2015). *Diversity in organizations: A critical examination*. New York, NY: Routledge.
36. DiTomaso, N., Post, C., & Parks-Yancy, R. (2007). Workforce diversity and inequality: Power, status, and numbers. *Annual Review of Sociology, 33,* 473–501.
37. Milliken, F. J., & Martins, L. L. (1996). Searching for common threads: Understanding the multiple effects of diversity in organizational groups. *Academy of Management Review, 21*(2), 402–433.
38. Swann, W. B., Polzer, J. T., Seyle, D. C., & Ko, S. J. (2004). Finding value in diversity: Verification of personal and social self-views in diverse groups. *Academy of Management Review, 29*(1), 9–27.

39. Thomas, R. R. (1990). From affirmative action to affirming diversity. *Harvard Business Review, 68*(2), 107–117.
40. van Dijk, H., van Engen, M., & Paauwe, J. (2012). Reframing the business case for diversity: A values and virtues perspective. *Journal of Business Ethics, 111*(1), 73–84.
41. Department for Business Innovation and Skills. (2013). *The business case for equality and diversity: A survey of the academic literature*. BIS Occasional Paper No. 4. Retrieved from https://www.gov.uk/government/publications/the-business-case-for-equality-and-diversity-a-survey-of-the-academic-literature
42. Kossek, E. E., Lobel, S. A., & Brown, J. (2006). Human resource strategies to manage workforce diversity: Examining the business case. In A. M. Konrad, P. Prasad, & J. K. Pringle (Eds.), *Handbook of workplace diversity* (pp. 53–74). London UK: Sage.
43. Jones, K. P., King, E. B., Nelson, J., Geller, D. S., & Bowes-Sperry, L. (2013). Beyond the business case: An ethical perspective of diversity training. *Human Resource Management, 52* (1), 55–74.
44. Ebie, S., & Djebarnis, R. (2011). Equality and diversity management: An investigation into the business case for equality and diversity practise in SMEs. *The International Journal of Diversity in Organisations, Communities and Nations, 10*(5), 145–167.
45. Delton, J. (2007). Before the EEOC: How management integrated the workplace. *Business History Review, 81*(2), 269–295.
46. Kossek, E. E., & Pichler, S. (2006). EEO and the management of diversity. In P. Boxell, J. Purchell, & P. O. Wright (Eds.), *Handbook of human resource management* (pp. 251–272). Oxford, UK: Oxford University Press.
47. Rutherglen, G. (2015). Private rights and private actions: The legacy of civil rights in the enforcement of Title VII. *Boston University Law Review, 95*(3), 733–757.
48. Plaut, V. C. (2010). Diversity science: Why and how difference makes a difference. *Psychological Inquiry, 21*(2), 77–99.
49. Plaut, V. C. (2010). Diversity science: Who needs it? *Psychological Inquiry, 21*(2), 168–174.
50. Podsiadlowski, A., Gröschke, D., Kogler, M., Springer, C., & van der Zee, K. (2013). Managing a culturally diverse workforce: Diversity perspectives in organizations. *International Journal of Intercultural Relations, 37*(2), 159–175.
51. Gotsis, G., & Kortezi, Z. (2015). Different approaches to managing a diverse workforce. In G. Gotsis & Z. Kortezi, *Critical studies in diversity management literature: A review and synthesis* (pp. 13–21). SpringerBriefs in Psychology. Dordrecht, Germany: Springer.
52. Prasad, P., & Mills, A. J. (1997). From showcase to shadow: Understanding the dilemmas of managing workplace diversity. In P. Prasad, A. J. Mills, M. B. Elmes, & A. Prasad (Eds.), *Managing the organizational melting pot: Dilemmas of workplace diversity* (pp. 3–30). Thousand Oaks, CA: Sage.
53. Kelly, E., & Dobbin, F. (1998). How affirmative action became diversity management. *American Behavioral Scientist, 41*(7), 960–985.
54. Fryberg, S. A., & Stephens, N. M. (2010). When the world is colorblind, American Indians are invisible: A diversity science approach. *Psychological Inquiry, 21*(2), 115–119.
55. Stryker, S. (1980). *Symbolic interactionism: A social structural version*. Menlo Park, CA: Benjamin Cummings.
56. Tajfel, H. (1978). The achievement of group differentiation. In H. Tajfel (Ed.), *Differentiation between social groups: Studies in the social psychology of intergroup relations* (pp. 77–98). London, UK: Academic Press.
57. Stets, J. A., & Burke, P. J. (2000). Identity theory and social identity theory. *Social Psychology Quarterly, 63*(2), 224–237.
58. Tajfel, H., & Turner, J. C. (1986). The social identity theory of intergroup behaviour. In S. Worchel & W. G. Austin (Eds.), *Psychology of intergroup relations* (2nd ed., pp. 7–24). Chicago, IL: Nelson-Hall.

59. Turner, J. C., Hogg, M. A., Oakes, P. J., Reicher, S. D., & Wetherell, M. S. (1987). *Rediscovering the social group: A self-categorization theory*. Oxford, UK: Blackwell.
60. Haslam, S. A., van Knippenberg, D., & Platow, M. J. (Eds.). (2014). *Social identity at work: Developing theory for organizational practice*. London, UK: Psychology Press.
61. Albert, S., & Whetten, D. A. (1985). Organizational identity. In L. L. Cummings and B. M. Staw (Eds.), *Research in organizational behavior. An annual series of analytical essays and critical reviews*, Vol. 7 (pp. 263–295). Greenwich, UK: JAI Press.
62. Gioia, D. A., Price, K. N., Hamilton, A. L., & Thomas, J. B. (2010). Forging an identity: An insider-outsider study of processes involved in the formation of organizational identity. *Administrative Science Quarterly, 55*(1), 1–46.
63. Corley, K. G. (2004). Defined by our strategy or our culture? Hierarchical differences in perceptions of organizational identity and change. *Human Relations, 57*(9), 1145–1178.
64. Pratt, M. G. (2012). Rethinking identity construction processes in organizations: Three questions to consider. In M. Schultz, S. Maguire, A. Langley, & H. Tsoukas (Eds.), *Constructing identity in and around organizations* (pp. 21–49). Oxford, UK: Oxford University Press.
65. Kreiner, G. E., Hollensbe, E., Sheep, M. L., Smith, B. R., & Kataria, N. (2015). Elasticity and the dialectic tensions of organizational identity: How can we hold together while we are pulling apart? *Academy of Management Journal, 58*(4), 981–1011.
66. van Knippenberg, D., & Sleebos, E. (2006). Organizational identification versus organizational commitment: Self-definition, social exchange, and job attitudes. *Journal of Organizational Behavior, 27*(5), 571–584.
67. Riketta, M. (2005). Organizational identification: A meta-analysis. *Journal of Vocational Behavior, 66*(2), 358–384.
68. Avery, D. R. (2011). Support for diversity in organizations: A theoretical exploration of its origins and offshoots. *Organizational Psychology Review, 1*(3), 239–256.
69. Shore, L. M., Randel, A. E., Chung, B. G., Dean, M. A., Ehrhart, K. H., & Singh, G. (2011). Inclusion and diversity in work groups: A review and model for future research. *Journal of Management, 37*(4), 1262–1289.
70. Lau, D. C., & Murnighan, J. (1998). Demographic diversity and fault lines: The compositional dynamics of organizational groups. *Academy of Management Review, 23*(2), 325–340.
71. Carton, A. M., & Cummings, J. N. (2012). A theory of subgroups in work teams. *Academy of Management Review, 37*(3), 441–470.
72. Chrobot-Mason, D., Ruderman, M. N., Weber, T. J., & Ernst, C. (2009). The challenge of leading on unstable ground: Triggers that activate social identity faultlines. *Human Relations, 62*(11), 1763–1794.
73. Héroux, S., & Fortin, A. (2016). Innovation: The influence of diversity and IT competence of boards of directors and executive management. *The International Journal of Organizational Innovation, 8*(3), 18–43.
74. Ruiz-Jimenez, J. M., & del Mar Fuentes-Fuentes, M. (2016). Management capabilities, innovation, and gender diversity in the top management team: An empirical analysis in technology-based SMEs. *Business Research Quarterly, 19*(2), 107–121.
75. Lindgren, M., & Packendorff, J. (2009). Social constructionism and entrepreneurship. *International Journal of Entrepreneurial Behaviour & Research, 15*(1), 25–47.
76. Levinson, H. (1971). Conflicts that plague family businesses. *Harvard Business Review, 49* (2), 90–98.
77. Hoffer Gittell, J., & Douglass, A. (2012). Relational bureaucracy: Structuring reciprocal relationships in roles. *Academy of Management Review, 37*(4), 709–733.

78. Uhlaner, L. M. (2008). *The role of ownership in governance: A neglected focus in entrepreneurship and management research*. Inaugural Lecture. Breukelen, The Netherlands: Nyenrode Business University. Retrieved from http://www.nyenrode.nl/FacultyResearch/research/Documents/Inaugural%20lectures/uhlaner_inaugural_lecture.pdf
79. European Commission. (2014). *Overview of diversity management implementation and impact amongst Diversity Charter signatories in the European Union*. European Commission—Directorate-General for Justice. Retrieved from http://ec.europa.eu/justice/discrimination/files/diversity_report2014_en.pdf

Gender and Entrepreneurship in Angola: Narratives of the 'Muambeiras' of Lubango

Evanice Nadya and Emilia Fernandes

Abstract The financial and economic situation of most African countries forces individuals to create their own jobs in order to generate some sort of income for themselves and their families. This chapter presents a study that aims to look into the entrepreneurial experiences of a particular group of female entrepreneurs from Lubango, Angola: the *Muambeiras*. After conducting interviews with a small group of *Muambeiras*, five life stories were analysed to illustrate the gender difficulties these women must overcome to do business, as well as the importance of business practice as a way of dealing with gender discrimination. This study contributes to a greater understanding of gender and entrepreneurship, through an account of the diversity of how women carry out entrepreneurship and how they deal with gender discrimination. Moreover, it contributes to redefine policies that can support these women's business activities and acknowledge their important roles in the community.

Introduction

In the last years, and in several countries, entrepreneurship has been greatly commented on, both in the media and by institutional government organisations [2, 26]. These discourses have presented entrepreneurship as a way of creating alternatives to employment and economic growth, as well as constituting an approach to dealing with financial crises [18, 25]. However, in northern and western countries, these discourses are not gender-neutral. Indeed, they are being associated to a certain masculinity, which is linked to an entrepreneurial practice perceived as exclusively economic, and connected to a psychological profile that is culturally and socially ascribed to men [1, 5].

E. Nadya · E. Fernandes (✉)
Department of Management, School of Economics and Management,
University of Minho, Campus Gualtar, 4710-057 Braga, Portugal
e-mail: mifernandes@eeg.uminho.pt

C. Machado and J.P. Davim (eds.), *Managing Organizational Diversity*,
DOI 10.1007/978-3-319-54925-5_6

In many African countries, entrepreneurship is associated to a form of economic and social survival, and not necessarily as a business opportunity [20]. The financial and economic situation of most African countries forces individuals to create their own jobs in order to generate some sort of income for themselves and their families. However, this type of entrepreneurship is often performed by women. As a social group, women have less access to the formal economy and employment. Due to war, as well as political and social reasons, they are often those who are chiefly responsible for their families, and are the only 'breadwinners' which families can count on [8]. One of the countries where these gendered contours can be found is Angola [9, 20].

This exploratory study aims to look into the entrepreneurial experiences of a particular group of female entrepreneurs from Lubango, Angola, who are identified as *Muambeiras*. These businesswomen import and commercialise diverse products, such as clothes, fashion accessories and furniture. The *Muambeiras* of Lubango play a major role in family economy, and are an important figure in the trading and economic activity of their communities.

After conducting interviews with a small group of *Muambeiras*, five life stories were analysed. These illustrate the gender difficulties they must overcome to do business, as well as the importance of business practice as a way dealing with gender discrimination and standing up against the subordinate position of women in Angolan society.

The study of this business experience undertaken by women, which has rarely been done before, will allow one to contribute to a greater understanding of gender and entrepreneurship, through an account of the diversity of how women carry out entrepreneurship and how they deal with gender discrimination. Moreover, this exploratory study will contribute to a definition of the policies that can support these women's business activities and acknowledge their important roles in the community.

Gender and Entrepreneurship

In northern and western geographies, entrepreneurship has been perceived by research and societies as an economic practice where men constitute the main reference [31]. Entrepreneurial practice is often exclusively associated to instrumental and economic values which make use of a predominantly neoliberal discourse [1, 35]. Moreover, the ideal entrepreneur subject is constructed by resorting to masculinity, which is centred on individual and psychological traits that value ambition, competitiveness, and risk [5]. The institutional support given to entrepreneurship evaluates entrepreneurial projects based on such normative masculine profiles [4]. Moreover, educational programmes on entrepreneurship are constructed so as to value the masculine discourse, which favours young male students who intend to pursue an entrepreneurial project [10, 33].

This gender and normative construction of entrepreneurship, and of the entrepreneur subject, creates social and cultural barriers for the entry and permanence of women in entrepreneurial practice. They are perceived as being less capable than male entrepreneurs [1], and their companies are attributed less value [6] or are related to specific types of business which are socially perceived as feminine. Female entrepreneurs experience greater difficulty in accessing financial investment [27]. The lack of gender equality when requesting institutional credit often makes female entrepreneurs depend on their family's financial support, which further reduces their opportunities of being autonomous in their businesses and in the decision-making process itself. Moreover, family tasks continue to be socially viewed as being women's responsibility [19, 22, 27]. However, female entrepreneurs often have to deal with family resistance to their entrepreneurial practice [15, 27]. All these barriers position women asymmetrically in relation to men and they are never totally recognised as "real" entrepreneurs.

In recent years, a line of research focused on gender and identity has explored how female entrepreneurs deal with gender discrimination and reclaim legitimacy in their business contexts. Female entrepreneurs have experienced a dual presence, in the sense that they are related to two different and contradictory meanings of gender [5]: the symbolic and social masculine meaning of entrepreneurial practice, and the symbolic and social feminine meaning associated to women. Several studies (e.g. [1, 5, 17, 19]) show how female entrepreneurs tend to move between these meanings in order to achieve and maintain their legitimacy at work and in their professional contexts as entrepreneurs. As an illustration, the study by Essers and her co-authors [15, 16] reflects on the life stories of immigrant female entrepreneurs and how they deal with ethnic constraints, gender and a normative masculine profile of entrepreneur. For these women, and in their particular ethnic context, this profile can be synonymous with a way of reclaiming their emancipation [15], or exploring and developing a new female ethnicity founded on gender equality [16]. Gherardi [19] studied the way in which female entrepreneurs negotiate and subvert the dichotomous gender discourse of the work-family life balance by challenging the assumptions that female work threatens family, and that women in entrepreneurship will be affected by the potential work-family imbalance, since reproduction and home are women's first and main responsibilities.

The constant increase in studies focused on gender and entrepreneurship has allowed one to understand the difficulties and experiences of female entrepreneurs from many different parts of the world (e.g. [12, 28, 32]). However, most literature tends to focus on northern and western countries. By choosing Angola and the life stories of the *Muambeiras* as an empirical context, this exploratory study aims to broaden the scope of the knowledge produced concerning the experiences of female entrepreneurs in an African context.

Female Entrepreneurs in Angola: The Case of the 'Muambeiras'

Entrepreneurship in the African context is often associated to economic survival and not to business opportunities as such [24]. Entrepreneurial practice is understood by individuals as a strategy to overcome high rates of unemployment and poverty [20]. This type of entrepreneurship is particularly associated to African women [20]. In Angola, women lead the numbers of entrepreneurship, with a rate of 25. 2% when compared to the 20.3% of men [9]. As in other African contexts, entrepreneurship emerges as a way for socially devalued groups to deal with poverty and obtain some economic income for their families. Usually, most of the entrepreneurial practice undertaken by women is situated outside the formal economy [13]. Informal markets play important social and economic roles in a country like Angola and coexist with the formal economy [14].

However, most of the entrepreneurial practice undertaken by women remains invisible and is not acknowledged by economic and political agents, institutions or Angolan society. Their voices are rarely considered when institutions discuss ways of dealing with the legal and cultural constraints of entrepreneurial practice, nor when polices are suggested to overcome such difficulties and promote entrepreneurship in this particularcontext [22]. In fact, the institutions of Angola have been considering several polices and measures to boost entrepreneurship and bring informal entrepreneurs into the formal economy. To achieve such a purpose, the Angolan government has developed several institutional programmes: the National Development Plan (Plano Nacional de Desenvolvimento—PND), Angola Invests (Angola Investe), programmes of exclusive attendance and for credit support to the entrepreneur (Guiché Único do Empreendedor—GUE; Balcão Único do Empreendedor—BUE), Programme of Young Angola (Programa Angola Jovem—PAJ). The main goal of these institutional programmes is to fight poverty and unemployment, since Angola recently experienced extremely violent armed conflict [7]. Both government and formal institutions have acknowledged the significance of entrepreneurship as a critical practice in aiding the sustained development of Angola, as well as a way of promoting structural changes [20]. However, within this set of government initiatives, there is no policy with the aim to promote equal gender opportunities in entrepreneurial practice or the recognition of the several women who are already involved in entrepreneurship. Female entrepreneurs in Angola remain invisible, and their social and economic roles in the Angolan communities are not recognised or legitimised by institutional power. Moreover, the few research studies which have focused on entrepreneurship in Angola have favoured an economic and gender-neutral approach to entrepreneurship. This has reinforced the invisibility of female Angolan entrepreneurs, even if international and national reports point to their highly significant participation in the informal economy. Angolan society has maintained a dominant and conservative discourse with regard to gender, where women are culturally and socially associated to family and reproduction [22].

Methodology

Context of the Study: The Entrepreneurial Practice of 'Muamba'

In the Angolan context, 'Muambeiras' is the name given to female entrepreneurs who import several different products from foreign countries, in order to commercialise these in the informal economy of their communities [13].

The name of 'Muambeiras', which these female entrepreneurs are known as, was attributed to them by the workers of the Luanda *Quatro de Fevereiro* International Airport [3]. In order to carry out their entrepreneurial practice, Muambeiras have to travel to foreign countries to look for the products they wish to commercialise in their communities. Several international markets constitute their trade routes; these include many Chinese cities, Dubai, Los Angeles, New York, Bangkok, Brazil, Turkey and Lisbon. It is in these countries that the Muambeiras buy many products with the intention of importing them to Angola, such as clothing, shoes, jewellery, hair extensions and other fashion accessories [13]. In this sense, they need to know which fashion trends will sell, as well as where the commercial and trade sources for their products lie. These are indispensable conditions for their success in Muamba [3, 34].

The entrepreneurial practice of Muamba implies that female entrepreneurs have to possess a certain degree of autonomy, which contradicts the values of a patriarchal society as is the case of Angola [34]. This is the reason why the work of the Muambeiras is not favourably regarded in Angola, since these women are often away from their families; they travel frequently and refuse to submit to a male family figure of authority [3].

Barreau-Tran [3, p. 3] describes these Angolan female entrepreneurs as being a group which is characterised by some diversity. Muambeiras can be female entrepreneurs from low social classes with low educational levels, owners of clothing and fashion accessories stores, market sellers and female professionals looking for other forms of economic income to supplement their low salaries. For many of these women, their entrepreneurial practice is a solution to the high rates of unemployment, and to the gender discrimination they experience in the formal economy [23].

As in other countries of Africa, Muambeiras play an important role in the economic survival of their families and communities [13]. The civil war experienced in Angola has resulted in women bearing great responsibilities in their families. Female entrepreneurs often live in a context of single-parent families, where the male element died due to civil war or the children are the result of pregnancy in adolescence [3, 29].

This exploratory study aims to analyse the gender experiences of a group of Muambeiras from Lubango, a city in the south of Angola. In particular, it analyses which types of gender difficulties Muambeiras have to overcome to do business, and how their business practice can be perceived as a way of dealing with gender discrimination in Angolan society.

Method of Research and Participants

Research concerning the gender experiences of Muambeiras is very scarce; it is for this reason that an exploratory approach was adopted. This exploratory study resorted to a qualitative methodology [11], using the semi-structured interview as a research method. Most of the interviews took place between January and February of 2015, at the *Muambeiras*' homes, and lasted an average time period of 52 min. Specific questions were asked regarding their personal trajectories, before they decided to be involved in *Muamba*, and the constraints and challenges that they are currently facing. Other questions were also asked concerning the advantages and disadvantages of being female and an entrepreneur at the same time, and their reflections about their contributions to the community in the position of female entrepreneurs. The 'snowballing' method was used to contact the *Muambeiras* [30]. The interviews were conducted with eight *Muambeiras* of the Lubango region. Most of these women occupied another professional position as public employees in the education and health sectors, while simultaneously exercising the Muamba practice.

The low number of interviews ensued from the fact that most of the Muambeiras who were contacted did not feel comfortable about sharing their stories; they were afraid that this could generate negative implications in their personal and business lives. It must be remembered that *Muambeiras* are predominantly related to an informal economy.

All the interviews were tape-recorded and transcribed. The qualitative data was gathered to elaborate the analytical categories about the *Muambeiras*' experiences. In the first phase of qualitative analysis, open coding was used to find common patterns and themes in the participants' stories. For this chapter, one selected the part of the *Muambeiras*' stories, where they talk about their gendered constraints on initiating *Muamba* and doing business, and the strategies used to overcome such gendered difficulties. A selection of the five narratives where these themes homes, and was undertaken.

The Life-Stories of the Lubango 'Muambeiras'

In this section, excerpts from the life stories voiced by five *Muambeiras* are presented. All the names used to identify the *Muambeiras* are pseudonymous. Catarina is 42-years old and finished high school. She has been involved in *Muamba* for 11 years although she also has a job in public service. She owns a fashion store in the city centre, and employs two female workers. Before she sold clothes and fashion accessories, she had tried to launch other types of business. She lives with her husband and six children. Vilma is 40-years old, has completed primary school, and sells women's clothes. She also simultaneously works as a public employee. She does not have a store to commercialise her products, and this forces her to receive her clients at home. She has been involved in Muamba for 10 years. Vilma is single and has two children, whom she leaves with her mother and niece. Before practising *Muamba*, Vilma sold home-made cookies, second-hand clothes, and alcoholic drinks. Julia is 37-years old, and terminated the second year of university. She is a nurse who then became involved in Muamba 16 years ago. She sells women's clothes from home. Julia is single and lives with her three children and her mother. Guida is 41-years old, has a degree in history, and is a public employee; she has been in the Muamba business for 16 years. Guida started her business by selling women's clothes and she also presently owns a furniture store, another food store, an ice-cream shop and a business in cleaning services, whose major clients are public institutions. She currently has nine employees working for her. She lives with her husband and four children. Doming as is 35-years old and finished primary school. She is a public employee and has practised *Muamba* for 9 years. She sells women's clothes from home. She is married and has three children.

Context and Constraints of Entrepreneurial Practice

In their discourses concerning entrepreneurial practice, female entrepreneurs shared an experience of economic hardship at a very young age, which left them with no other choice but to work, even when they were children. Family needs were stated as the first reason for starting to work at such an early age. Vilma confessed that her family lived in very poorly conditions, which were aggravated by the Angolan civil war.

> When I was a child, we experienced very difficult moments. We arrived in Lubango as refugees. My parents were unemployed and we could not study because we didn't have money. At the age of 9, I started to help in the kitchen. I began selling cookies at 5 a.m. in the market, and at 9 a.m. I went to school. The money that I got from this allowed me to help my family and to pay my meals at school (…) At the age of 12, I sewed children's clothes, which I sold in the market. We had no ready-made clothes as we see today. Until my twenties, I lived from this work.

They also shared several experiences of precarious work before becoming involved in *Muamba*. Like Vilma, Catarina also had many work experiences before starting to commercialise imported products.

> I tried different jobs. I even worked in television as an assistant but I didn't have a salary, only some money to help my family and me. I thought that it would be better to look for something different (…). I also sold ice-creams, which I made at my parents' home when I was younger. When I grew up, my responsibilities become greater because I got married and I tried to find other ways of doing business. I sold bread. At that time, there was only me doing that and two other bakeries in the city, which were not enough to meet people's needs. I also sold meat.

This diversified business experience is also shared by Domingas.

> I sold bread, afterwards cookies, ice-cream. I also sold cigarettes and clothes to my neighbours.

Some of these experiences are, in fact, related to a practice of entrepreneurship which allowed the participants to contribute economically to their families, first their parents and siblings and, afterwards, their own families. Like Vilma, Domingas talked about the need to help her mother and siblings to make some money, so as to allow her family to surpass poverty.

> We had many financial difficulties and that's why I had to enter the business world very early. My mother's salary was too low, and for that reason I started selling soup in the Maringa Square (…). With that money I bought school material for my brothers and me.

Julia emphasised the need to have another form of income besides her salary as a public employee.

> It is impossible to live only on my salary. Getting into business and afterwards becoming involved in Muamba was a way of bringing money to my family, and allowing my children to go to school. I have a family to take care of and I am the main person who is t responsible for them.

Family is an important key-factor in the discourse of female entrepreneurs. Family, along with the need to escape poverty, is the reason that leads these women to progressive autonomy and the will to overcome the gender constraints which women in Angolan society have to deal with. Family is, in fact, the justification women use to start their entrepreneurial practice; this has allowed them to exercise greater power in their families, since they are the major or first economic contributor to the family. In this sense, *Muambeiras* often become the main breadwinners in their families. By occupying such a position, they are thus transgressing the conservative gender norms which attributed such a position to men.

However, the family also symbolises the conservative values inherent to the role of women in Angolan society, and this constitutes an important constraint for women when starting their entrepreneurial practice as *Muambeiras*. Catarina shared that her entrepreneurial activity was not always supported by her husband and, in order to go on her first business trip, she had to run away from home.

> My first business trip is a very long story. I had to ask for my husband's permission to travel to Brazil and I was not allowed. So I had to run away. I remember that I only took one piece of clothing, a dress, with me and I lied saying that I was going to Windhoek because, at that time, we flew Lubando-Windhoek-Johannesburg-São Paulo. I went with my cousin. When I arrived, I sold everything that I had brought with me. After that, my husband never told me that I was forbidden to travel or do other things.

Julia also shared that she lied to her family about her first trip to the USA because she was scared that they would forbid her to go.

> I went alone to the USA and when I went on that trip I did not tell my family the truth. I knew that they would forbid me to go because the journey was very long and I would be going alone.

These women's constant need to travel is not positively viewed by their families and communities. It implies long periods of time away from home and the total transgression of the gender power relations in their communities, where women are perceived as dependent and submissive to the male family figures [21, 22].

Moreover, travel implies that these women will (re)appropriate the meanings and values considered to be part of the male symbolic universe [1, 5]. In a globalized world, they are thus able to define themselves: through freedom of movement, economic autonomy, and the willingness to take risks. Concerning these trips, Catarina emphasised the freedom and the adventurous nature of her business trips.

> I really love to travel. I started with Brazil, and when the prices got too high in this country, I went to China. I have been in this business for 11 years and I thought that I should know and explore other countries and cities, and that is why I started traveling to Miami, New York, and Lisbon.

Most of these female entrepreneurs invested in a journey without having detailed information about the countries they visited during their business trips, and without knowing anything about the language or cultures of those countries. Nonetheless, they mentioned the help of some local guides, so that they reveal an awareness of the high risk involved in going on these business trips. However, the general feeling is one of experiencing an adventure, testing their limits, and being able to surpass themselves and their fears. Julia told us that her first journey to the USA was an adventure, since she went to this country without any support or knowledge about the country's markets.

> At that time, Muambeiras didn't usually travel to the USA with Emirates (the travel company), and I decided to go via Dubai. I was the only African person on the flight from Dubai to the USA! During the trip, I was always praying to God to help me to finish my journey, to be brave, and not to be scared. Alone on the flight and alone in the city!

Vilma also felt very proud of herself when she made the decision to explore the furniture markets in China, and decided to invest in interior home decoration.

> I went to China alone, I did not go with anyone, and I did not have a local guide (…) 90% of the things I brought and got were a result of my research and curiosity.(…); when you're curious, you start being very attentive to what is happening around you, comparing prices and different materials. I did all that alone, with God's blessing.

Julia also expressed her need to take constant risks and explore new markets.

> In the nineties, I went to Namibia to buy clothes and sold them here in Lubango (…); afterwards, I started to explore Congo Brazzaville, in particular Pointe-Noire. We go where we think it will be profitable. Then I started to explore Brazil, where I still go very often (…) I also went to China but the markets did not have anything that had to do with me and with my business.

Years of experience bring greater know-how of the business, which then becomes an expertise in the global markets. One can see this from the way Guida talks about the markets she explores.

> In Brazil, clothes are of high quality but much more expensive. In China, we have cheap products but they do not invest in quality. However, the products of high quality in China are more accessible than those in Brazil. In fact, many things we bought in Brazil were made in China. Today I am in the catering service business, and I always go to China to buy the material and equipment that I need for my work.

The risk inherent to these women's entrepreneurial practice is also present in the insecurity they said that they felt at airport customs. Corruption is also present in their daily lives, when they have to receive their luggage at the airport. The fact that they are women seems to create a higher probability of being harassed by the customs airport staff, who always tend to change the taxes they have to pay and make some of their luggage disappear; the airport police often humiliate them by searching them, when they are treated as potential drug dealers.

> At Windhoek airport we saw that all our luggage had been opened and searched. It was something that we could always expect from the airport staff. During our most recent trips we, the Muambeiras, always had a bus at the airport door waiting for us; it took us to the hospital for a check-up to see if we had drugs in our bodies. I wonder if they would do that if we were men. We were only released at the end of the day. Meanwhile, they opened our bags, stole some of our things. We could only realise that when we arrived home, with no chances of making any complaints (…). The truth is that, when we complain, nothing happens (Catarina)

Julia complained about the Luanda airport and the corruption they have to deal with.

> All the Muambeiras know how difficult it is to arrive at our airport. We have experienced very sad and stressful situations created by the customs staff, who always ask for more money to pay our country's taxes. We are very badly treated by these people, and we know what they are doing, but we can say nothing if we want to continue carrying out our business.

The country's corruption and economic instability influence the *Muambeiras*' lives. Their poor image, which is related to their gender condition and their need to continue doing business, places these women in very complicated situations.

Access to credit is also another problem for the *Muambeiras*. The gender of the *Muambeiras* seems to be linked to the idea that banks see their businesses as not being "real businesses". Moreover, these women often operate in an informal economy, which further complicates their access to credit. This is Guida's case; she has often tried to obtain credit but has never managed to do so because she was not considered to be a true entrepreneur.

> In the beginning, they (the bank personnel) say that everything is ok, that they do not need any more information or documents. Then, time passes and they never give us an answer. In the end, they say that there is something missing and that our project is not a really good project, but they never explain why. I think they have never intended to give me credit, because they do not see me as a real businessperson.

The *Muambeiras* thus experience the prejudice of doing business as *Muambeiras* and of being women doing business. Their families and communities seem to accept their entrepreneurial practice because the relation of dependence has somehow been transgressed: families need the economic income these women generate, and communities need their businesses in order to have access to imported products with some credit facilities, which the *Muambeiras* tend to provide to their clients. However, this acceptance does not solve all the difficulties they face. In symbolic terms, *Muambeiras* still have a poor image in the community and are considered to be transgressing their gender roles.

Strategies to Overcome Gendered Difficulties and Entrepreneurial Identity

Due to the submissive role of women in Angolan society, and to the fact that Muamba often belongs to an informal economy, *Muambeiras* are always exposed to great risk and find themselves in a rather weak position, which they need to manage on a daily basis. As a result of all this, the *Muambeiras*' discourses show that they are always dependent on the relationships of trust they create with others, namely with other *Muambeiras*. The *Muambeiras*' network is a constant topic in their discourse; it constitutes a strategy used to deal with the difficulties that they have to overcome in their business practice. This network is particular important for *Muambeiras* when they travel to foreign countries. Julia stressed how important it was for her to find more *Muambeiras* in the hotel that she stayed in, in the USA, during a business trip.

> I was so happy when I found out that there were more Angolan Muambeiras staying in my hotel. They came to introduce themselves to me, one after the other (…) Later, I made friends with one of them. They explained that this trade route was very profitable, despite the great distance they have to cover. They also told me that I should go to Los Angeles, and gave me directions to get there: do this, and then that, and so and so (…). In China I found Fatima and I liked her. We were in the same conditions: both alone and lost. So we decided to trust each other and started shopping together, trading together. When I went to Dubai with my cousin, I took Fatima with me, and Fatima also buys things for me when I cannot travel.

Catarina said that it was very important for her to meet a *Muambeira* whom she admired for being so successful and independent. Domingas, on the other hand, mentioned that she always travels with other female *Muambeira* friends with whom she shares a room in the hotels where they stay, in order to reduce expenses. Domingas also uses the contacts established through her *Muambeira* friends to do business.

> During one of my trips through Angola, I met a Muambeira, who I became friends with, and she told me about the Brazilian hair business (…). I went on my first international trip with that friend and she helped me to buy Brazilian hair, which was a great success among my clients (…) Muambeiras were traveling to China very often and, due to my friend's suggestion, I also decided to invest in this market (…). I did not like the Chinese markets because the products were of low quality. However, on my return trip, I talked to other Muambeiras and they told me about the Turkish markets.

Muambeiras also talked about the importance of trust relationships with their clients, who are often also women. Julia said that she preferred to have older women as clients because they are more responsible and always pay for what they buy.

> I have clients that have been with me for a long time, and these clients always bring new clients whom I can trust. I often give my clients credit and I need to trust them. Young girls also pay their debts but, because of a few that did not pay, I prefer the older female clients.

The experience of having to sell their products from home makes these trust relationships extremely important. By having a shop, Catarina was able to overcome such difficulties. She believes that clients who do not pay only do so because they do not see *Muambeiras* as real businesspeople.

> I have clients that have owed me money for 10 years. They don't take my work seriously. Sometimes it is very difficult to get the money and we cannot complain. The problem is more serious when we sell from home; clients tend to see us as friends and some forget to pay, or delay the payment of the products they take for a long time.

However, they all agreed that their success in sales, whether in a shop or at home, depends on their trust relations with most of their clients.

Their success as entrepreneurs is often related to their conquest of freedom, to their children's well-being, and to economic profit. In Carolina's case, such success implies values of beauty and elegance, which is partly connected to a certain cultural construction of femininity.

> I hope to continue to practise Muamba because I love my work. I need to travel, to be in contact with other people, and to visit other types of stores, where you can see beautiful clothes and shoes. This is my life.

The female construction of the *Muamba* business is also present in Julia's discourse, when she reflects on her contribution to the community, and to the emancipation of other women.

> I help to modernise the image of our city. I have given credit to my female clients who, thanks to my products, can dress better and feel better about themselves. I see how they get satisfaction from that. I see how we broaden women's horizons.

Female entrepreneurs also emphasise their relation with other cultures and customs, as well as how they learn to understand *difference* through such experiences. Domingas speaks about that experience and how it has allowed her to understand the importance of feeling free to make her own decisions, of being a self-confident woman.

> Contact with other cultures, and ways of being, has made me think about my life, about what I want for myself and for my business. It has helped me to educate my sons; but it has also allowed me to become surer of my value, and more independent.

Vilma considers herself to be an important person in her community because of her involvement in Muamba. In her view, no one can deny that she is a highly successful and independent businesswoman.

> I am very proud of myself. In fact, I am one of most important people here in my community in terms of doing business. I do not depend on my ex-husband, I was able to feed my siblings, pay for their school, and am now responsible for my own family.

Guida emphasises that the Lubango community has to value *Muambeiras* because they bring innovation to the city, and they create jobs when the business starts to grow.

> I do not work alone at present. I have 9 permanent employees with me. I have improved my life but also that of the people who work with me.

This feeling of pride and achievement is also present in the way these women see the future. Those that do not have a store to sell their products, hope to achieve such an aim in the future. That is Julia's case.

> My dream is to have my own shop. Now I am still working as a public employee and I cannot travel as often as I wish. When I have a shop, I can hire people who can help me to go further.

The discourse of risk is also included in the construction of the participants as entrepreneurial subjects. Guida resorts to the meaning of risk to define her way of doing business, and that has changed over time.

> Nowadays, I know that risk is part of my job. I have to see further, to increase the amount of money I invest in the products I import (…). I have to consider that I have more competition than I had a few years ago as well.

Discussion and Conclusion

In this chapter, the life stories of a group of *Muambeiras* from Lubango was explored. The aim of this exploratory study was to analyse how these female entrepreneurs deal with gender when they do business, and how business can contribute to the emancipation of their gender. We demonstrated how family and community contexts affect their life choices, and how the context produces an indirect effect on their very early stages of economic independence. This indirect effect has helped these women to become emancipated from their families and to reclaim legitimacy in their communities. As in other African countries [20, 24], women are often their family's main support and the main breadwinners. However, they still occupy a gender position of submission in relation to men. They are underestimated in the formal economy and have less access to employment [20–22]. *Muamba* arises as a professional and work alternative in order to deal with poverty, unemployment, low salaries, and the lack of job opportunities.

This exploratory study also shows that most of the difficulties which these women deal with are related to their country's political and economic conditions: corruption, economic instability, and difficulties in accessing credit. However, their gender seems to aggravate these difficulties. The combination of being women as well as *Muambeiras* generates contradictory effects in the lives of this group of female entrepreneurs. They are constrained and discriminated against in their entrepreneurial practice, not only because they are women but also due to the fact that they have obtained their emancipation through that same practice.

Moreover, the strategies inherent to their business practice help *Muambeiras* to overcome some of the gendered social and economic constraints. The most important of these seems to be the existence of an informal network, which promotes affective relations and work support among *Muambeiras*. This female network allows women to share information about trade routes, to travel together to foreign countries, and to create ties of friendship amongst themselves.

Muamba is also a practice that allows these women to construct a positive image of themselves by resorting to meanings related with freedom of movement, independence, ambition, a willingness to take risks, and adventure. The fact that the *Muambeiras* reclaim such meanings implies that they end up being representatives of a male ideal of the construction of the entrepreneur subject, which is often reinforced by mainstream research and society [5, 31]. Through this identity of entrepreneurial work [15], the women transgress and symbolically escape from their submissive position and from the restrictive female meanings associated to women in Angolan society [21, 22]. Moreover, they appropriate a certain degree of masculinity to define what it means to be a *Muambeira*, without attaching such meanings to men. In this sense, the reclaiming of such an identity, and the attribution of value to the *Muamba* practice, has allowed these women to be progressively recognised and legitimised in their communities as business subjects, and thus enter the formal economy.

To conclude, this exploratory study contributes to an understanding of the complicated but acknowledged reality of female entrepreneurs by researching a context which has seldom been explored. One thus hopes to bring greater visibility to the experiences of a particular group of female entrepreneurs, who play an important role in the economy and social life of Lubango. Such knowledge can aid the Angolan Government, as well as public institutions, to define specific policies to promote equal opportunities in gender-related work. This would support the recognition of *Muambeiras* in their communities, and would allow them to continue with their businesses, namely by facilitating access to financial investment, and by intensifying the ethical control of the airport customs services.

References

1. Ahl, H. (2006). Why research on women entrepreneurs needs new directions. *Entrepreneurship: Theory and Practice, 30*(5), 95–621.
2. Ahl, H., & Nelson, T. (2015). How policy positions women entrepreneurs: A comparative analysis of state discourse in Sweden and the United States. *Journal of Business Venturing, 30*(2), 273–291.
3. Barreau-Tran, L. (2013). Trajetos e Trajetórias de comerciantes angolanas na cidade de São Paulo. IV Colóquio Internacional de Doutorandos do CES, 6–7 dezembro, cabo dos trabalhos.

4. Brindley, C. (2005). Barriers to women achieving their entrepreneurial potential. *International Journal of Entrepreneurial Behavior & Research, 11*(2), 144–161.
5. Bruni, A., Gherardi, S., & Poggio, B. (2004). Entrepreneur-mentality, gender and the study of women entrepreneurs. *Journal of Organizational Change Management, 17*(3), 256–268.
6. Brush, C., & Gatewood, E. (2008). Women growing businesses: Clearing the hurdles. *Business Horizons, 51*(3), 175–179.
7. Catessamo, F., & Rua, O. L. (2015). Inovação e empreendedorismo em Angola: Contribuições para o desempenho das PME. Porto: vida económica editorial SA.
8. Chant, S. (2014). Exploring the "feminisation of poverty" in relation to women's work and home-based enterprise in slums of the Global South. *International Journal of Gender and Entrepreneurship, 6*(3), 296–316.
9. CEIC. (2008). Estudo sobre o empreendedorismo em Angola. Centro de Estudos e Investigação Científica da Universidade Católica de Angola.
10. Costa, A. S. M., & Saraiva, L. A. S. (2012). Hegemonic discourses on entrepreneurship as an ideological mechanism for the reproduction of capital. *Organization, 19*(5), 587–614.
11. Denzin, N. K., & Lincoln, Y. S. (1994). Introduction: Entering the field of qualitative research. In N. K. Denzin & Y. S. Lincoln (Eds.), *Handbook of qualitative research* (pp. 1–17). Thousand Oaks: Sage Publications.
12. Díaz-García, M.-C., & Welter, F. (2011). Gender identities and practices: Interpreting women entrepreneurs' transitions. *International Small Business Journal, 31*(4), 348–404.
13. Ducados, H., & Ferreira, M. (1998). O Financiamento informal e as estratégias de sobrevivência económica das mulheres em Angola: A Kixikila no município do Sambizanga – Luanda. *Documentos de trabalho - CEsA, 53,* 1–14.
14. DUM, J. E. E. (2014). Mercados Informais do Município Sumbe da Província do Kwanza Sul: Uma evidência empírica. *Tese de Mestrado em Contabilidade e Finanças*. Instituto Politécnico de Bragança.
15. Essers, C., & Benschop, Y. (2007). Entrepreneurial identities: Female entrepreneurs of Moroccan or Turkish origin in the Netherlands. *Organization Studies, 28*(1), 49–69.
16. Essers, C., Benschop, Y., & Doorewaard, H. (2010). Female ethnicity: Understanding Muslim Immigrant businesswomen in the Netherlands. *Gender, Work and Organization, 17*(3), 320–339.
17. Fenwick, T. J. (2002). Lady, inc.: Women learning, negotiating subjectivity in entrepreneurial discourses. *International Journal of Lifelong Education, 21*(2), 162–177.
18. Fernandes, E., & Leite, R. (2014). The gender of the entrepreneur and the politics of entrepreneurship for women: an exploratory study. In C. Machado & J. P. Davim (Eds.), *Work organization and human resource management* (pp. 105–123). Switzerland: Springer.
19. Gherardi, S. (2015). Authoring the female entrepreneur while talking the discourse of work-family life balance. *International Small Business Journal, 33*(6), 649–666.
20. GEM—Global Entrepreneurship Monitor. (2012). Estudo sobre o empreendedorismo em Angola. GEM.
21. Grassi, M. (2001). Género, desigualdades sociais e desenvolvimento na África subsaariana: o caso de Angola. *Afriche e Orienti, 1,* 1–17.
22. Grassi, M. (1998). O Papel da mulher empresária Angolana no desenvolvimento do país: empresárias de Luanda e Benguela. *Economia Global e Gestão, 3*(1/2), 209–226.
23. Governo de Unidade e Reconciliação Nacional, GURN (2000). Relatório de seguimento das metas da cimeira mundial pela infância. Governo de Unidade e Reconciliação Nacional.
24. Hechavarria, D. M., & Reynolds, P. D. (2009). Cultural norms & business start-ups: The impact of national values on opportunity and necessity entrepreneurs. *International Entrepreneurship and Management Journal, 5,* 417–437.
25. Kenny, K., & Scriver, S. (2012). Dangerously empty? Hegemony and the construction of the Irish entrepreneur. *Organization, 19*(5), 615–633.
26. Krefting, L. A. (2002). Re-presenting women executives: Valorization and devalorization in US business press. *Women in Management Review, 17*(3/4), 104–119.

27. Loscocco, K. A., & Robinson, J. (1991). Barriers to women's small business success in the United States. *Gender & Society, 5,* 511–532.
28. Maden, C. (2015). A gendered lens on entrepreneurship: Women entrepreneurship in Turkey. *Gender in Management: An International Journal, 30*(4), 312–331.
29. Medina, M. (2001). Direito da família. Luanda: Colecção da Faculdade de Direito da Universidade Agostinho Neto.
30. Noy, C. (2008). Sampling Knowledge: The hermeneutics of snowball sampling in qualitative research. *International Journal of Social Research Methodology, 11*(4), 327–344.
31. Ogbor, J. O. (2000). Mythicizing and reification in entrepreneurial discourse: Ideology-critique of entrepreneurial studies. *Journal of Management Studies, 37*(5), 605–635.
32. Pettersson, K., & Cassel, S. H. (2014). Women tourism entrepreneurs: Doing gender on farms in Sweden. *Gender in Management: An International Journal, 29*(8), 487–504.
33. Petridou, E., Sarri, A., & Kyrgidou, L. P. (2009). Entrepreneurship education in higher educational institutions: The gender dimension. *Gender in Management: An International Journal, 24*(4), 286–309.
34. Silva, E. (2011). Tradição e Identidade de género em Angola: Ser mulher no mundo rural. *Revista Angolana de Sociologia, 8,* 21–33.
35. Tedmanson, D., Verduyn, K., Essers, C., & Gartner, W. (2012). Critical perspectives in entrepreneurship research. *Organization, 19*(5), 531–541.

Expatriate Women: A Dream Waiting to Come True

Claudia Carvalho and Carolina Feliciana Machado

Abstract In a context of a globalized economy, business is no longer limited by national boundaries. To maintain their competitiveness, organizations are increasingly establishing and strengthening their presence overseas. For this reason, and to accomplish their projects more successfully, many firms take the option to deploy employees from the home's company to their subsidiary. This way, we have inevitably assisted to a strong rise of these professionals in international assignments. However, this rise has been very disproportional to men and women, with a clear disadvantage do women. Indeed, women represent a clear minority compared to men. ***What explains such a huge difference that sets apart men and women?*** The present critical literature review starts from this issue in order to obtain a greater understanding of the gender effect in the process and in the experience of the expatriation (one of the many international forms of mobility in the organizational contexts).

Introduction

In a context of a globalized economy, in order to stay competitive and, in some situations, to guarantee its survival, organizations are, more and more, *spreading its wings* as well as expanding its businesses across borders. For this reason, and in order to realize its projects with great success, worker's displacement from the home's company to other international contexts where they operate has been increasingly an option to these organizations. Inevitably, we have been witnessing to a strongly grown of professionals in international missions. However, in a great disproportionately way for men and women, with a clear disadvantage for women. Indeed, women represent a minority in these worker's total universe. ***What explains such a huge difference that sets apart men and women?*** This present chapter of

C. Carvalho · C.F. Machado (✉)
Department of Management, School of Economics and Management,
University of Minho, Campus Gualtar, 4710-057 Braga, Portugal
e-mail: carolina@eeg.uminho.pt

C. Machado and J.P. Davim (eds.), *Managing Organizational Diversity*,
DOI 10.1007/978-3-319-54925-5_7

critical literature review starts from this issue with the aim to understand gender effect in the expatriation process and experience (one of the possible ways of international mobility in the organizational context). To this end, we will present, in a first step, the expatriation concept, exploring some statistical data about the same. Then, we are going to focus the concepts of gender (male and female) and gender stereotypes. Clarified the main concepts that guide this chapter, we will move to the core of this critical review, in other words, the relationship between expatriation and gender. The discussion will be focused in three specific issues, namely, the different perspectives about women role in the expatriation process; the factors that explain women under–use in these international assignments; and the change that had occurred over the last years.

Expatriation: Concept and Statistical Data

According to the Priberam Dictionary of Portuguese Literature [1] expatriation means "to send out motherland" and "Go out of their motherland to live abroad." By this side expatriate is "that one who lives, willingly or not, abroad its motherland." This concept analysis, further removed from scientific research, is important in order to understand the broad sense of expatriation, reason why different authors use the concept in a very distinct way. Indeed, this concept scientific use changes greatly, not only due to the fact but also the process had been started by the Government, by an organization or by the person itself, as well as on the basis of being or not a voluntary process [2].

To the scope of this work, we will consider the expatriate organizational vision which define him/her as the worker who, continuing to perform duties in his/her organization, is temporarily transferred to another country (by an established period of time), accompanied by their family [3–5]. It is important to highlight that, even when the expatriation focus is limited to this level, we find a variety of visions about the concept, essentially in what concerns the expatriate period of time. Many authors (e.g., [6]) understand expatriation as that who occurs until 5 years. Many others also define a minimum period of time, usually from 6 months to 2 years (e.g., [4, 7]). Moreover, Bonache et al. [8], Banai and Harry [9] and Fenwick [10] draw attention to most varied forms of mobility which go beyond the expatriation conventional vision that include workers who travel regularly to another country, not implying necessarily moving to it. However, it is important to point out that, due to its distinctive characteristics, some authors consider that these new mobility ways cannot be included in the great expatriation cluster [5].

To better understand the expatriation process, we must bear in mind that it involves three phases: pre-expatriation; expatriation and repatriation [11, 12]. The first phase assumes an assessment of the need of expatriation, including the entire recruitment and selection process as well as the applicant's preparation. The second phase refers to the international mission development. Finally, the third phase is related to the expatriate return to the home organization. Regarding to this last phase, it is important to clarify that expatriation implies that the worker contractual relation remains after the international mission and, once finished, the worker shall have the right to return to the home organization to a task to be defined, which should be compatible with the repatriated worker profile [13].

Expatriation has been used by organizations with a strategic character, as a way of more easily reach a well succeed internationalization process. Indeed, research has shown, in an extensive way, the expatriation potential to the internationalization success, by a diversified set of reasons among which we can highlight the need to anchor the subsidiary organization in the home organization policies and values (e.g., [5, 9, 14, 15]). Maybe by this strategic character, expatriation has been greatly associated to management positions. Indeed, expatriation begun to be confined to this type of positions, initially very associated to diplomats and professionals with military career, whose professional nature required to stay out of the country [16]. However, nowadays expatriation includes a great number of professionals [17], as it is confirmed by data in the "Global Relocation Trends: 2012 Survey Report" where it is revealed, inclusively, that the number of expatriate workers in technical tasks overpass those that are in management tasks [18]. The option by workers expatriation with a profile highly technic is usually relate with the difficulty in finding this type of worker locally (in the host country) [13].

Given the relevance of internationalization and expatriation to organizations, this is a phenomenon in great expansion. According to the study developed in 2011 by Pricewaterhouse Coopers, international assignments increased 25% in the last decade, forecasting an increase of 50% until 2020. At the Portuguese national level AEP [14] forecast that expatriation can increase about 20%. The study "International Assignment Survey—Portugal 2010" developed by Mercer Consultancy, reinforce the increase trend of workers' expatriation processes of the Portuguese organizations.

Gender and Gender Stereotype: Concepts

Defined the expatriation concept, it is critical to clarify the concept of gender, frequently confounded with that of sex (female or male). When we refer to one' person sex, we are pointing out the basic differences that exist between men and women in a physiologic way [19]. By this side gender is the identity—female or male—that each of us interiorize and which include specific behavioral roles and standards (Oakley 1972 referred by [20]). This is, so, a sociocultural construction; a construction that, as Delphy [20] highlights is imbued of an intrinsic asymmetry and hierarchy that split groups. This happens because "They are assumptions, however, that have tended to be developed and refined in contexts dominated by males and, hence, have been disadvantageous to females" [21, p. 352]. So, in a context characterized by a traditional male domain, women were during centuries associated to the private/domestic/familiar spherefamiliar sphereprivate spherepublic sphere and men to the public/productive/of work spherework sphere [21]. It is from these roles internalization that result female and male identity female identity [22], the first associated to a great sensibility and emotionality and the second to a great independence, aggressiveness and dominance [23].

This woman and man image become, this way, a "kind of prototype that come into being as a norm to a team" [19, p. 10]. Thus, often, people form an opinion about others characteristics and behaviors based only in their gender—female or male [24, 25]. This type of beliefs about different attributes and roles of men and women (gender stereotypes) are widely shared and are extremely embedded in societies [25, 26]. Izraeli et al. [26] distinguish three types of gender stereotypes—*gender characteristics stereotype, gender role stereotype, and gender professions label.*

Gender characteristics stereotype refers to the embedded beliefs about differences in personality traits between men and women [26]. As noted before, women are considered as being more emotional, dependent, conformist, passives, and less ambitious and rational; the opposite is valid to men [23, 26].

Gender role stereotype concerns to beliefs related to behaviors considered appropriated to men and women [26]. This type of stereotype is based in beliefs such as those that defend that women primary responsibility is to look after children; women should not give orders to men, and men should not receive orders from women [21, 26]. While the first type of stereotypes (*gender characteristics*) has a particular descriptive character—how men and women *are*—; this second type of stereotypes has a prescriptive component—how men and women *should behave* [26].

Finally, *gender professions label* is associated to the expectation that a profession is more suitable to a gender than to the other (e.g., police should be men and nurses should be women) [26].

Despite not having, necessarily, a correspondence to reality, these differences have been understood as natural, "as being within individuals," in other words, as a result of the physiological differences [25, p. 12]. This narrative was built over centuries by the own science and fueled by media, even in the face of subsequent scientific evidences that do not support it (for instance, [27]). In addition, a number of empirical studies have demonstrated that men and women are not quite so opposites and that there exist more similarities than discrepancies, and that when they exist "are usually small and based in the group medium behavior" [25, p. 16]. However, this is not a thematic free of controversy, considering that are, also, various the studies that still focus differences between man and women (for instance, [28]).

The first aim of stereotypes "is that of simplify and organize a complex social environment, becoming it less ambiguous. But they also interest to justify team discrimination and originate biases" [25, p. 13]. In the case of gender stereotypes, once they convey beliefs of superiority and male dominance (and in the reverse of the medal, of female inferiority and submission), these can originate preconceptions related to womenwomen, and so, to discrimination acts that look to harm and impair women exclusively because they belong to the "female group" [25]. These acts can assume a way, more or less explicit, more or less subtle [26]. At the professional world they have been, over time, strongly hard to women [24], as we will have the opportunity to explore in the following pages.

Although by definition stereotypes are persistent and stables, research as demonstrated that these are not immutable. Indeed, it seems to be possible to change stereotypes, namely through the simple exposition or contact with persons with atypical characteristics to the stereotype [29, 30].

Although by definition stereotypes are relevant and stable, research has demonstrated that they are not immutable. Indeed, it seems to be possible the stereotypes change namely through the simple exposition or contact with people with atypical characteristics to stereotype [29, 30].

Expatriation and Gender: "*Who Should We Choose?*" and "*Who Is Really Chosen?*"

In a context with a strong tendency in the expatriation increase (as showed above), of course that one of the first questions that researchers and human resource managers immediately face is "*Who should we choose?*". The answer to this issue is not, far from, linear. However, studies have demonstrated that this choice should attend not only to the hard kills (namely, technical professional competences), but also to the soft skills (namely, personal competences such as the adaptation to new cultures/realities, the ability to new challenges/flexibility/ability to adapt to the change) (e.g., [5, 13, 31]). Moreover, literature has pointed out to the need to not neglect some contextual variables (namely, ***The candidate profile matches the host country and the specific business sector characteristics?***) (e.g., [32]), as well as familiar issues (***What is the family situation?, Are there family difficulties not possible to be managed by the organization?***) (e.g., [13]).

Related to this last issue, but not less important than this one, researchers have, also, looked to answer to the inevitable issue: "***Who is really chosen?***". If we take into account the set of selection factors previously announced, the answer will be variable, although it becomes important to refer that scientific research has pointed out to a generalized tendency under which organizations select their collaborators based in the experience and the previous performance of the organization [31, 33]. However, when we consider only gender variable, this is frighteningly uniform. Indeed, in what concerns gender, literature points out, in unison, to the answer "Men" [24, 34–37]. While it is true that the growing women participation in the workforce has been one of the most striking features of the global labor market since the latter half of the twentieth century [38], it is no less true that it has been lacking a women full inclusion and that their participation has been limited.

Nancy Adler, one of the main researchers that has dedicated her studies to the understanding of the gender inequalities in the international management area, wrote in 1984 [39] a pioneer article entitled "*Women in international management: Where are they?*". That time, this author studies pointed to the existence of only 3% of expatriate women. ***More than 30 years after what have changed in expatriates selection?*** The most recent research has revealed that too little. We verify, from a consistent way, the persistence of a women subrepresentation in these international assignments. Indeed, and according the literature review developed by Altman and Shortland [40], although it appears some changes, women representation in expatriate processes is still very low. Reviewed studies identify variable percentages; however, never exceeding 15%.

The own language used in the literature about international management reinforce this disparity. Indeed, in the scientific context it is deeply used the concept of "expatriate," except when exists a clear intention in making reference to women in this situation. In this case, the "expatriate"expatriate concept is replace by the "expatriate woman" concept, highlighting her status of outsider, as in the case of, for instance, "firefighter woman" and "police woman" [41]. This inequality is such relevant that has deserved the edition of specialized journals in gender studies in organizations, such as, "*Women in Management Review*" and "*Gender, Work and Organization.*"

In addition, it is important to highlight, henceforth, that gender inequality observed in the expatriate process is not an isolated reality. Indeed, inequality between men and women is felt in a transversal way in different areas, not only in profession, but also in health, education, politic, among others [25]. At the professional level, disparities go well beyond expatriation. We highlight, among others, women difficulties in having access to management jobs, specially, leadership positions, where we observe a male predominance [42], as well the gap in compensation received by men and women, significantly lowest in the second group [43, 44].

Expatriation and Gender: "*Chose We Well?*"

Although expatriation contains within itself a success potential, not always international assignments follow the ideal course (e.g., [45, 46]). Expatriation process failure might result not only in the premature return, as well as, and according to some authors view, in less tangible ways, namely at the attitudinal and behavioral level [47–49] behavioral level. Some research [48, 49] have been highlighting to the fact according to what failure could not be so high as initially considered (some studies pointed out to early return rates around 40%). Notwithstanding, although numbers cannot be as distressing, they must be valued; specially if we consider that an unsuccessful expatriation process implies high costs (and not only economic) to the organization and the expatriate [5].

Of course that, looking to the high costs of an unsuccessful expatriation process, we immediately face the following question: ***But, what has caused this failure in the expatriation process?*** The answer seems to lie particularly, in the expatriate and/or his family intercultural adaption difficulties (when a join displacement happens), but also in family problems that appear when the family is separated (when the family does not go to the another country jointly with the expatriate) (e.g., [50–52]). According to Harris [36, p. 181]. "The most significant feature of the research into expatriate failure rates is that it is based on a mate population. This research contains actual evidence of male expatriates facing cross-cultural adjustment problems and family problems. In contrast, research conducted on the outcome of women's global assignments indicates that female expatriates are successful in their assignments." ***What kind of evidence supports this statement? Are women effectively more qualified to this kind of assignments?***

Answer to these questions is not consensual being shrouded in discussion. The idea of "female capacities" to missions have received, simultaneously, an intense support and opposition, as further we can observe in regard to the general belief that women and men are completely different, as we have focused earlier.

The concept of "female capacities" have been supported in a considerable number of studies that, on the one hand, have shown the personality characteristics relevance to the expatriation success and, on the other hand, have revealed gender differences in these characteristics. In 2000 [53] Caligiuri, in her literature review, had found effectively strong support arguments to the personality role in the expatriate adaptation and success. However, she had also verified the inexistence of validation studies that support these arguments. So, the author had decided to develop a validation study in which had collected data related to the well know personality traits "Big Five" (also known as McCrae and Costa 5 factor model) and to the performance of 143 expatriates in mission in 25 different countries. According to her results, expatriates that had a highest rank in extroversion and kindness factors showed a less desire in finish their mission early; in addition, the performance of those that showed higher conscientiousness, was assessed more favorably by their supervisors. Based in this knowledge, as well as in studies that support personality differences between men and women, it started to emerge, which Altman and Shortland [40] in their literature review call of, "challenging speech," with many voices advocating the possibility according to which women are more appropriate to international assignments and others strongly claiming it (e.g., [24, 28, 36, 38, 54, 55]). On the whole, these studies have revealed that women show best interpersonal competences than men, namely in what concerns sensitivity, flexibility, empathy, and sociability. Guthrie et al. [28] begin their article with the title "*Are women 'better' than men?*" and conclude that women have personality traits (extroversion, conscientiousness, and kindness) that make them naturally better candidates. These "female characteristics" seem to be associated to a greater ability for women to interact and establish alliances with other expatriates and the local community, favoring this way their adaptation and the assignment success [38, 53].

Altman and Shortland [54] compare women in international assignments to "aliens" that became masters in the art of survival in a foreign environment strongly dominated by men. The "alien" concept runs from a resilience capability of these women who manage to be successful as well, in more adverse conditions than men. Indeed, studies have showed that, even in the most hostile environments, expatriate women are so successful than men (e.g., [38, 56, 57]). In their literature review, Altman and Shortland [40, p. 210] pointed out that "Mounting recent evidence confirms that women adapt better than men in cross-cultural business situations."

But, why are women subject to more adverse conditions than men? It has been observed that expatriate women are effectively exposed to more challenges than men, both at professional and personal level [58]. According to many studies, women not only are strongly subject to discrimination from their expatriate colleagues [26, 59], but they also have to deal with more family conflicts, have yet the primary responsibility to take care of the child in an atypical and challenging situation [60]. All this is exacerbated by the fact that these women can feel that they are receiving less organizational support than their male colleagues [61, 62], at the same time that they have the "need" to be extraordinary competent in order to reach the recognition of their work [56, 63]. Indeed, it has been verified that, since women belong to a minority on the expatriates whole (phenomenon described by Kanter, in 1977, that has designated it by *tokenism*), they must be exceptionally competent to obtain the majority acceptance [24].

Therefore, since they were better prepared to deal with adversity, and that their presence in international assignments origins a great positive impact in local relationships, it has been defend that women are the *ideal* candidates to international assignments [54].

Not all authors agree, however, with this type of position. To the authors that have criticized this perspective, these studies are feeding a superior concept, concept against which these same authors have protested in relation to men and that now defend in what concerns women. Hofbauer and Fischlmayr [64] are two of the authors that present a strong confrontation to the "female abilities" concept, arguing that this concept suggests these abilities universality and does not have in attention social diversity, both female and male. These authors do not call into question the great emphasis of women relational abilities identified in the study; what they inquire is that "Given the evidence about women showing high amounts of those skills, we still ask whether it makes sense to characterize them as 'female,' that is, to assign a set of skills to the entire group of women and thereby suggest that other types of skills characterize men only" [64, p. 49]. The authors challenge this universalistic and essentialist concept of gender characteristics, supported in Kanter research, particularly when they highlight that "To put it in more general terms, Kanter holds that features that we address as gender specific do, in fact, arise from certain patterns of conduct that emerge under certain situational circumstances" (Kanter 1977, referred by [64, p. 51]). For instance, power conditions held by a person can promote or hinder certain attitudes or behaviors. Thus, it is no logic, to these authors, and in line with that showed by others regarding to other areas of strong debate about gender, such as leadership (e.g., [65]), to consider women more or less appropriate to international assignments. An individual characteristics and capabilities presumption cannot, in the light of this perspective, be based only in its gender. It must also consider their life and work experiences, their professional position and their training. Moreover, Wajcman [65] and Hofbauer and Fischlmayr [64] consider that the focus in "individual female characteristics" draw attention away from structural barriers that they face. Recognize the value of "female qualities" in management and in international assignments is important but not necessarily sufficient so that asymmetries can be correct and that more women can be included in these positions. On the other hand, these authors consider that it is possible that, while a still dominant force, men can hold this "female" rediscovery and include it in the traditional "male repertory". Thereby, adding new qualities to those that, allegedly, they already possess, men will be in advantage and women will continue to be seen as endowed only with the traditional "female characteristics". This focus in the differences could, this way, contribute to an organizations' reinforcement in traditional gender stereotypes. Authors advocate, alternatively, the need to a focus in any stereotype challenge, positive or negative, relative to women, or men.

In her article "*Global managers: no longer men alone*" [66], Adler directs the debate about differences between men and women to the idea of diversity and complementarity between both. Although focused in differences in management skills between men and women, the general idea of this perspective add value to the discussion and could eventually, be applied to expatriation (we cannot forget that many expatriates held management positions). The author defends differences between men and women; however, considers it imperative that they cannot be understood in the light of any value judgement. In other words, both men and women have strengths (as well as weaknesses), either innate or socially constrained. However, these differences must be view complementary, i.e., as contributing synergistically to the all, without a superiority or inferiority character being assigned to them. In a word characterized by a strong globalization and competitiveness, and according Adler vision, other authors have pointed out to the strong contribution that women can offer to organizations [67, 68], not because they are, in any way, "superiors," but because they add greater diversity, diversity this that has been pointed has a source of innovation and competitive advantage [24, 34].

Although the different perspectives addressed have a different focus, they have in common the recognition of the need of women be considered by organizations at the moment when they decide to recruit collaborators to the expatriation process. Alluding, thus, to the issues that has served as theme to Harris article [36, p. 175, 177]: "Why are women not selected for international management assignments? Why, therefore, do organizations continue to under-use such a valuable source of diversity and a potentially powerful aid towards developing a truly global mindset?" Next chapter looks to help to clarify these questions.

Expatriation and Gender: *Why Are Women Less Chosen Than Men?*

In recent years research has been addressing in an extensive way on the reasons to women be consistently underrepresented in the expatriation processes.

Regardless the different explanatory factors that have been presented, it is crucial contextualize, since now, these gender inequalities in our History. Indeed, we must be keep in mind that these inequalities are deeply rooted in the men and women historical journey, and in the different roles traditionally given to them, as focused earlier in relation to the concept of gender. In this context, it has been consistently verified the existence of a number of barriers (discriminative) which limit women progression and access to leadership positions and international experiences (and, particularly, to the participation in expatriation processes). These barriers power is such that, metaphorically, to report to the same, authors often use concepts like *glass ceiling* (Hymowitz and Schellhardt 1986, referred by [69]), *glass border* (Mandelker's 1994, referred by [70]), and *expatriate glass ceiling* [71].

Understanding these differences roots on the choice of men and women to expatriation, we must now list and analyze some of the specific motives that have been present in the literature.

Result of an extensive body of work, Adler had identified three factors that help understanding the female shortage in international assignments and that the author denominate as "myths" [34, 56, 72]: (1) "Women do not want to be international managers," (2) "Companies refuse to send women overseas," (3) "Foreigner's prejudice against women renders them ineffective, even when interested and sent" [56, p. 176]. These myths were test by a wide research body in order to check their correspondence with reality.

In what concerns the first myth, some data pointed out that women interest is at the same level of that of their male colleagues (e.g., [36, 72, 73]. However, there are some contradictory results that suggest gender differences in expatriation interest. According to Lowe [74] cultural distance and the host country development will be at the origin of a less female interest in some reference countries.

With regard to the second myth, there are strong arguments that lead us to presume that it has correspondence with reality [57, 73, 75–78]. It seems to be a persistent bias in women selection process based in gender stereotypes, assuming, for instance, that they are less available due to family reasons, they do not have the necessary skills suitable with the expatriate profile, at the same time that potential risks as well as the prejudice against women in host countries is also mentioned [36, 72, 73, 75].

Finally, relative to the third myth, in 1987 Adler [56] verified that there is no base to the women fear of discrimination in host countries. The author verified that these countries nationals' attitudes to these foreign women are different from their attitudes regarding to local women. What happens is that the condition of foreign prevails subject to the condition of woman, so that in this case women are judge by different standards from those of local women. Thus, it is not expected from foreign women the same behavior of local women. This phenomenon was designate by *Gaijin* syndrome (Japanese word that means "foreign"). Foreign woman is, this way, first seen as foreign, not as woman. These results were replicate in many research papers (e.g., [36, 77, 79, 80]). There is thus a great support to the idea that being woman is not necessarily a problem in host countries. There are also evidences that not only this is not a problem, as can represent an advantage, namely by the supposed easiness of woman in establish more positive relationships [53, 80].

In an alternative explanation, some authors have also considered these women influence to their scarcity in the international management. Inadvertently, they can self-induce barriers that can difficult their access to international experiences [61, 71, 81]. At this point, literature highlights to the fact of woman behave herself according to gender stereotyped expectative, assuming a passive role, revealing low self-confidence, relegating to a second level their career perspectives rather than their husbands/partners. These barriers are, according to Altman and Shortland [40] reinforced, among others, by an absence of social and organizational support and a weak organizational networking.

Expatriation and Gender: *What Changes?*

Over time, different authors have done their forecasts relative to the evolution of the women presence and role in the expatriation processes. Some of them with optimistic perspectives about the women future in these international assignments, believing in the barriers overcoming (e.g., [36, 54, 66, 82]), others with more doubts and a greater criticism (e.g., [61, 64]). Whereas it is undeniable that women sub representation remains the same, this is also undeniable the increase (even when reduced) that had occurred since Adler [39] research about 30 years ago. ***What has caused this change? Will we be moving to a gradual but effective and deep change? Will we be assisting to a barriers' removal that have blocked the women international presence?***

Has we had the opportunity to exploit previously, according to some studies, women present in a greater degree interpersonal and intercultural competences indispensable to a well-succeed international assignment [28, 40]. For this reason, Adler [34, 66] defend that, in order to remain competitive, organizations are forced to overcome gender stereotypes and to recruit according rational criteria, such as qualifications and competences. Only in that way will be possible to achieve diversity that will guarantee the necessary innovation and distinctiveness. In the same line of thought, Altman and Shortland [40] consider that this increase may be due to the claimed ability that women present to deal with adversity, and therefore become better candidates to this kind of assignments. Thus, and according to this position, through a greater awareness of the women role in international markets, organizations have provided the required lever to change.

Other authors (e.g., [61, 71, 75, 76]) have placed the main emphasis, not in the organizational thinking change, but in the individual thinking and behavior change. Fischlmayr [61] notes, in her study, the relevance of women to assume a proactive attitude in their own international mobility. While the majority of expatriate men of her study were approach by the organization in order to engage in these assignments, the majority of women needed to apply their transfer. However, the author also shows that these women are an exception to the rule, claiming that the majority continue to take on a passive role. ***It will not, however, reasonable, to assume that we will assist increasingly to a change in women relation with work?*** Altman and Shortland [40] answer in the affirmative to this question. In their literature review, the authors draw the attention to the further women guidance to career. Indeed, there are strong evidences that new generations have more ambitious women, with professional ambitions similar to men, willing to give up their family and personal life to the detriment of their career shaping and that are opposed actively to the discriminatory barriers that are put upon them. It is possible, thus, that if this trend continues (which has been consistently consolidated over time), women are the key players of their own growth in the international paths.

Hofbauer and Fischlmayr [64] present a different reinterpretation of this tendency. In their critical analysis study, the authors conclude that: "Thus women entering the domain of international management may turn out to be the conquerors of 'empty castles,' as men actually deserted the terrain and moved on to more prestigious positions in international organizations" [64, p. 48]. Effectively, an extensive range of empirical studies (e.g., [48, 57, 63, 70, 76, 83]) have revealed that it is common women occupy lower hierarchical positions compared to those of men.

Moreover, it has been argued that international assignments are becoming less attractive to men who are, thus, less available to accept them, "forcing" organizations to extend ever wider its access to women [40, 64]. Traditionally, the collaborators participation in international assignments has been see as a lever to the career development and the access to leadership positions [5, 84]. ***What discourage, therefore, collaborators to join to the expatriation processes?*** The answer to this question lies in the potential adverse effects of expatriation, which have become increasingly emphasize, raising a cloud over the "glamour" formerly associated with this type of experiences.

First of all, although some contradictory results exist (e.g., [85]) on the whole it has been verified that participation in expatriation processes does not imply necessarily changes in professional development, at least not in the short term [54, 86–90]. Indeed, it has been consistently verified that skills developed by expatriates during assignments are underutilized by organizations when they come home; collaborators are not integrated in new positions suited to the new skills; and they do not receive compensation superior to those received by their no' expatriate colleagues. Hofbauer and Fischlmayr [64] claim, therefore, that given the less connection of international assignments to career progression, men opt to stay at the parent company, believing that "near the sight, near the heart," they will have more hypothesis in developing their career [64].

As Benson and Pattie [86] emphasize, however, research about these experiences long-term impact have shown that individuals with significant international careers occupy positions hierarchically superiors; are more frequently promoted; and have higher compensation. Notwithstanding, it is possible that nowadays, in a context of crisis, this is not so true and that, anyway, short-term effect become the most visible to the no' expatriate colleagues, acting as a deterrent of their interest in these processes. ***However, and as Hofbauer and Fischlmayr*** [64] ***also questioned, would not it be reasonable to assume that if this question is present to men, will not it be even more to women?*** Effectively it seems to be. Selmer and Leung [63] verified in their study that, even having equal position and previous experience in expatriation processes, women tend to be reintegrate in lower hierarchical positions.

Thus, should not be women feel even less enticed to participate in this kind of international assignment than men? Not necessary. According to the literature review developed by Altman and Shortland [40] expatriation becomes less interesting to men but not to women who sustain their interest.

Can we assume that short-term results at the expatriates' career level are sufficient to discourage male collaborators (once interested) to search these experiences but not women? Will be women particularly focused in long-term effects? Several studies (e.g., [85, 86, 90]) have been demonstrate that collaborators that accept to participate in expatriation processes develop positive expectations about the experience impact. Even when they return home, and facing an inadequate integration of their organization, some studies have verified that these collaborators still maintain optimistic relatively to the experience results, although in the long term and in a different organization [86, 90]. ***Will have women a great desire in access to these international experiences? Will be women making use of this gap that was open to them to claim their role in the organizations and assume more prestigious positions (even though they may not be so much than formerly and that the impact is less than the obtained by men and not immediate)?***

Furthermore, we must bear in mind that the collaborators that accept the expatriation process are not only tempted by the leverage that the professional experience can give to their career (e.g., [91, 92]). Personal challenge and learning opportunities also appear to be key factors [90]. ***Will be women more motivated for reasons other than directly associated to the career development?***

Lesser attractiveness of international experiences has also been associated to major collaborator concerns with safety issues (since the September 11 attacks) and with the stress associated to the repatriation in their family and, in particularly, with their children educational and social development [93]. ***Once again, these concerns will not arise also to women?*** Some studies have verified that expatriate women still correspond to a very selective and restrictive group of women that are not married [48, 57] and are, usually, young (under 35 years of age) [48], responding to that that we have verified concerning women that assume management positions [40, 76], and contrasting with the verified concerning men. It is possible that this kind of questions do not arise in the same way to these youngest and single women.

Finally, the importance depreciation given to the expatriation processes has been associated to the change trend of their term structure. Shorter assignments have become increasingly frequent [94, 95], which, while on the one hand, represent an advantage to the organization (essentially because allow them a cost reduction) [96], on the other hand can represent a disadvantage to the collaborators as they decrease their impact in the professional development [97]. Indeed, as they are shorter, there are a decrease of the depth of the cultural exhibition in these assignments and, therefore, a much more limited competence development [97]. These assignments can, therefore, lost its relevance at the men eyes. ***At the same time that they are depreciated by men, will not it be possible that this kind of assignments become more attractive to women who will be able, therefore, to manage more easier their family life?*** This is one more issue that needs to be explored.

Some Final Remarks

Inequality between men and women is a presence so significant in our society that has justified the establishment, at the national and international level, of innumerous legal diplomas, Ministers and Commissions that seek to alleviate the differences between these two groups. Despite all the efforts, the dream of equality in the different spheres of the women life remains to be done. It is a "phenomenon" in such a way rooted, that has persisted over time, crossing generations, resisting to the diverse social efforts and circumventing legal obligations. Following the analysis conducted over the course of 9 years in 142 countries, and based in the found course, the World Economic Forum [44] conclude that *maybe* we can reach gender equality at work in 2095. At best, they will therefore occur 79 years until the dream can be achieve. Researchers and human resource managers have a critical role in this process of change, and cannot ignore their role. In what concerns research, it is necessary, however, that this start moving its attention from women that are missing in the expatriation process, to those that are in these processes as well as in other international assignments. It is necessary to deeply understand how human resource managers and the own women can contribute to an adequate management of the expatriation process.

References

1. Dicionário Priberam da Língua Portuguesa, retrieved from http://www.priberam.pt/dlpo/expatria%C3%A7%C3%A3o. Accessed in June 2015.
2. Norwood, D., & Stai, B. (2011). Expatriation and international business. *Allied Academies International Conference, 11*(1), 29–34.
3. Baruch, Y., Dickmann, M., Altman, Y., & Bournois, F. (2013). Exploring international work: Types and dimensions of global careers. *The International Journal of Human Resource Management, 24*(12), 2369–2393.
4. Caligiuri, P. M. (2000). Selecting expatriates for personality characteristics: A moderating effect of personality. *Management International Review, 40*(1), 61–80.
5. Rego, A., & Cunha, M. P. (2009). *Manual de gestão transcultural de recursos humanos* (1ª ed.). Lisboa: Editora RH.
6. Burckhard, A., & Stai, B. (2011). Research in expatriation. *Allied Academies International Conference, 11*(1), 1–6.
7. Mayerhofer, H., Hartmann, L. C., Michelitsch-Riedl, G., & Kollinger, I. (2004). Flexpatriate assignments: A neglected issue in global staffing. *The International Journal of Human Resource Management, 15*(8), 1371–1389.
8. Bonache, J., Brewster, C., Suutari, V., & De Saá, P. (2010). Expatriation: Traditional criticism and international careers—Introducing the special issue. *Thunderbird International Business Review, 52*(4), 263–274.
9. Banai, M., & Harry, W. (2004). Boundaryless global careers: The international itinerants. *International Studies of Management & Organization, 34*(3), 96–120.
10. Fenwick, M. (2004). On international assignment: Is expatriation the only way to go? *Asia Pacific Journal of Human Resources, 42*(3), 365–377.

11. Ashamalla, M. (1998). International human resource management practices: The Challenge of expatriation. *Competitiveness Review: An International Business Journal, 8*(2), 54–65.
12. Cerdin, J., & Pargneux, Le. (2009). Career and international assignment fit: Toward an integrative model of success. *Human Resource Management, 48*(1), 5–25.
13. Câmara, P. B. (2011). *A expatriação em Portugal*. Lisboa: RH Editora.
14. Associação Empresarial de Portugal—AEP. (2013). Nómadas globais. Impactos multiculturais na internacionalização. Retrieved from http://scholar.google.com/scholar?hl=en&btnG=Search&q=intitle:No+Title#0, accessed in June 2015.
15. Tungli, Z., & Peiperl, M. (2009). Expatriate practices in German, Japanese, U.K., and U.S. multinational companies.: A comparative survey of changes. *Human Resource Management, 48*(1), 153–171.
16. Freitas, M. (2000). *Como vivem os executivos expatriados e as suas famílias?* Núcleo de Pesquisas e Investigações, (1–117). Retrieved from http://hdl.handle.net/10438/3221. Accessed in June 2015.
17. Bonache, J. (2005). Job satisfaction among expatriates, repatriates and domestic employees. *Personnel Review, 34*(1), 110–124.
18. Brookfield. (2012). *Global Relocation Trends Survey*. Brookfield Global Relocation Services. Retrieved from https://espritgloballearning.com/wp-content/uploads/2011/03/2012-Brookfield-Global-Relocations-Trends-Survey.pdf. Accessed in June 2015.
19. Nogueira, C. (2001). Feminismo e discurso do género na psicologia social. *Psicologia e Sociedade., 13*(1), 107–128.
20. Delphy, C. (1993). Rethinking sex and gender. *Women's Studies International Forum, 16*(1), 1–9.
21. Mills, A. (1988). Organization, gender and culture. *Organization Studies, 9*(3), 351–369.
22. Lorenzi-Cioldi, F. (1994). *Les androgynes*. Paris: Presses Universitaires de France.
23. Powell, G. N. (1993). *Women and men in management*. London: Sage.
24. Caligiuri, P., & Cascio, W. (1998). Can we send her there? Maximizing the success of western women on global assignments. *Journal of World Business, 33*(4), 394–416.
25. Nogueira, C., & Saavedra, L. (2007). Estereótipos de género: conhecer para os transformar. In *A Dimensão de género nos produtos educativos multimédia* (pp. 11–30). Lisboa: Ministério da Educação—Direção Geral de Inovação e Desenvolvimento Curricular.
26. Izraeli, D. N., Banai, M., & Zeira, Y. (1980). Women executives in MNC subsidiaries. *California Management Review, 23*(1), 53–63.
27. Jacklin, C. N., & Baker, L. A. (1993). Early gender development. In S. Oskamp & M. Costanzo (Eds.), *Gender issues in contemporary society*. London: Sage Publications.
28. Guthrie, J. P., Ash, R. A., & Stevens, C. D. (2003). Are women 'better' than men? Personality differences and expatriate selection. *Journal of Managerial Psychology, 18*(3), 229–243.
29. Zajonc, R. B. (1968). Attitudinal effects of mere exposure. *Journal of Personality and Social Psychology Monograph Supplement, 9*(2), 1–27.
30. Amir, Y. (1969). Contact hypothesis in ethnic relations. *Psychological Bulletin, 71*(5), 319–342.
31. Cheng, H. L., & Lin, C. Y. Y. (2009). Do as the large enterprises do? *International Business Review, 18*(1), 60–75.
32. Causin, G. F., Ayoun, B., & Moreo, P. (2011). Expatriation in the hotel industry: An exploratory study of management skills and cultural training. *International Journal of Contemporary Hospitality Management, 23*(7), 886–901.
33. Harvey, M., & Novicevic, M. M. (2001). Selecting expatriates for increasingly complex global assignments. *Career Development International, 6*(2), 69–87.
34. Adler, N. J. (1994). Competitive frontiers: Women managing across borders. *Journal of Management Development, 13*(2), 24–41.
35. Culpan, O., & Wright, G. (2002). Woman abroad: Getting the best results from women managers. *International Journal of Human Resource Management, 13*(5), 784–801.

36. Harris, H. (2002). Think international manager, think male: Why are women not selected for international management assignments? *Thunderbird International Business Review, 44*(2), 175–203.
37. Selmer, J. (2001). Expatriate selection: Back to basics? *The International Journal of Human Resource Management, 12*(8), 1219–1233.
38. Caligiuri, P. M., & Tung, R. L. (1999). Comparing the success of male and female expatriates from a US-based multinational company. *International Journal of Human Resource Management, 10*(5), 763–782.
39. Adler, N. J. (1984). Women in international management: Where are they? *California Management Review, 26*(4), 78–90.
40. Altman, Y., & Shortland, S. (2008). Women and international assignments: Taking stock—A 25-year review. *Human Resource Management, 47*(2), 199–216.
41. Berry, D. P., & Bell, M. P. (2012). "Expatriates": Gender, race and class distinctions in international management. *Gender, Work and Organization, 19*(1), 10–28.
42. Comissão Europeia (2013). *Women and men in leadership positions in the European Union.* Publications Office European Commission. Brussels.
43. Comissão para a Igualdade no Trabalho e no Emprego. (2015). *Disparidades salariais entre homens e mulheres em Portugal.* Retrieved from http://www.cite.gov.pt/pt/acite/disparidadessalariais_05.html. Accessed in June 2015.
44. World Economic Forum. (2014). *The global gender gap report 2014.* Retrieved from http://www3.weforum.org/docs/GGGR14/GGGR_CompleteReport_2014.pdf. Accessed in June 2015.
45. Baruch, Y., Steele, D. J., & Quantrill, G. A. (2002). Management of expatriation and repatriation for novice global player. *International Journal of Manpower, 23*(7), 659–671.
46. Stroh, L. K. (1995). Predicting turnover among repatriates: Can organizations affect retention rates? *The International Journal of Human Resource Management, 6*(2), 443–456.
47. Bozionelos, N. (2009). Expatriation outside the boundaries of the multinational corporation: A study with expatriate nurses in Saudi Arabia. *Human Resource Management, 48*(1), 111–134.
48. Forster, N. (1997). The persistent myth of high expatriate failure rates: A reappraisal. *The International Journal of Human Resource Management, 8*(4), 414–433.
49. Harzing, A. W., & Christensen, C. (2004). Expatriate failure: Time to abandon the concept? *Career Development International, 9*(7), 616–626.
50. Agard, J. Y., & Mevel-Pla, N. (2011). Adjustment and expatriation of managers: Switching the places of socialization. *Global Management Journal, 3*(1), 62–80.
51. Caligiuri, P., & Tarique, I. (2006). International assignee selection and cross-cultural training and development. In I. Bjorkman & G. K. Stahl (Eds.), *Handbook of research on international human resource management* (pp. 302–322). England: Edward Elgar.
52. Harrison, D., & Shaffer, M. (2005). Mapping the criterion space for expatriate success: Task and relationship-base performance, effort and adaptation. *The International Journal of Human Resource Management, 16*(8), 1454–1474.
53. Caligiuri, P. M. (2000). The big five personality characteristics as predictors of expatriates? Desire to terminate the assignment and supervisor-rated performance. *Personnel Psychology, 53*(1), 67–88.
54. Altman, Y., & Shortland, S. (2001). Women, aliens and international assignments. *Women in Management Review, 16*(3), 141–145.
55. Westwood, R. I., & Leung, S. M. (1994). The female expatriate manager experience: Coping with gender and culture. *International Studies of Management and Organization, 24*(3), 64–85.
56. Adler, N. J. (1987). Pacific basin managers: A gaijin, not a woman. *Human Resource Management, 26*(2), 169–191.
57. Forster, N. (1999). Another "glass ceiling"?: The experiences of women professionals and managers on international assignments". *Gender, Work & Organization, 6*(2), 79–90.

58. Mayrhofer, W., & Scullion, H. (2002). Female expatriates in international business: Empirical evidence from the German clothing Industry. *International Journal of Human Resource Management, 13*(5), 815–836.
59. Stone, R. J. (1991). Expatriate selection and failure. *Human Resource Planning, 14*(1), 9–18.
60. Fischlmayr, I. C., & Kollinger, I. (2010). Work-life balance—A neglected issue among Austrian female expatriates. *International Journal of Human Resource Management, 21*(4), 455–487.
61. Fischlmayr, I. C. (2002). Female self-perception as barrier to international careers? *International Journal of Human Resource Management, 13*(5), 773–783.
62. Hutchings, K., French, E., & Hatcher, T. (2008). Lament of the ignored expatriate: An examination of organizational and social network support for female expatriates in China. *Equal Opportunities International, 27*(4), 372–391.
63. Selmer, J., & Leung, A. S. M. (2003). Expatriate career intentions of women on foreign assignments and their adjustment. *Journal of Managerial Psychology, 18*(3), 244–258.
64. Hofbauer, J., & Fischlmayr, I. (2004). Feminization of international assignments: Conquering empty castles? *International Studies of Management & Organization, 34*(3), 46–67.
65. Wajcman, J. (1996). Desperately seeking differences: Is management style gendered? *British Journal of Industrial Relations, 34*(3), 333–349.
66. Adler, N. J. (2002). Global managers: No longer men alone. *International Journal of Human Resource Management, Women in Global Business, 13*(5), 743–761.
67. Peteraf, M. (1993). The cornerstones of competitive advantage: A resource-based view. *Strategic Management Journal, 14*(3), 179–191.
68. Porter, M. E. (1992). *The competitive advantage of nations*. London: MacMillan.
69. Eagly, A. H., & Carli, L. L. (2007). Women and the labyrinth of leadership. *Harvard Business Review, 85*(9), 62–71.
70. Linehan, M., & Walsh, J. S. (1999). Senior female international managers: Breaking the glass border. *Women in Management Review, 14*(7), 264–272.
71. Insch, G. S., McIntyre, N., & Napier, K. N. (2008). The expatriate glass ceiling: The second layer of glass. *Journal of Business Ethics, 83*(1), 19–28.
72. Adler, N. J. (1984). Women do not want international careers: And other myths about international management. *Organizational Dynamics, 13*(2), 66–79.
73. Stroh, L., Varma, A., & Valy-Durbinc (2000). Why are women left at home: Are they unwilling to go on international assignments? *Journal of World Business, 35*(3), 241–255.
74. Lowe, K., Downes, M., & Kroeck, K. (1999). The impact of gender and location on the willingness to accept overseas assignments. *The International Journal of Human Resource Management, 10*(2), 223–224.
75. Kollinger, I. (2005). Women and expatriate work opportunities in Austrian organizations. *The International Journal of Human Resource Management, 16*(7), 1243–1260.
76. Linehan, M., & Scullion, H. (2001). European female expatriate careers: Critical success factors. *Journal of European Industrial Training, 25*(8), 392–418.
77. Paik, Y., & Vabce, C. (2002). Evidence of back-home selection bias against US female expatriates. *Women in Management Review, 17*(2), 68–79.
78. Vance, C., Paik, Y., & White, J. (2006). Tracking bias against the selection of female expatriates: Implications and opportunities for business education. *Thunderbird International Business Review, 48*(6), 823–842.
79. Tung, R. (1998). American expatriates abroad: From neophytes to cosmopolitans. *Journal of World Business, 33*(2), 125–144.
80. Varma, A., Toh, S., & Budhwar, P. (2006). A new perspective on the female expatriate experience: The role of host country national categorization. *Journal of World Business, 41*(2), 112–120.
81. Linehan, M., & Walsh, J. S. (2000). Work-family conflict and the senior female international manager. *British Journal of Management, 11,* S49–S58.

82. Tung, R. (2005). Can women succeed as global managers? Executive Education, *China Europe International Business School, 5*, 1–3. Retrieved from http://www.ceibs.edu/pdf/execed/e.inspire1.womensc,pdf. Accessed in June 2015.
83. Linehan, M., & Walsh, S. (2001). Key issues in the senior female international career move: A qualitative study in a European context. *British Journal of Management, 12*(1), 85–95.
84. Silva, C., Carter, N., & Beninger, A. (2012). *Good intentions, imperfect execution? Women get fewer of the "hot jobs" needed to advance*. New York: Catalyst.
85. Suutari, V., & Brewster, C. (2003). Repatriation: Empirical evidence from a longitudinal study of careers and expectations among Finnish expatriates. *International Journal of Human Resource Management, 14*(7), 1132–1151.
86. Benson, G., & Pattie, M. (2008). Is expatriation good for my career? The impact of expatriate assignments on perceived and actual career outcomes. *The International Journal of Human Resource Management, 19*(9), 1636–1653.
87. Lazarova, M., & Caligiuri, P. (2001). Retaining repatriates: The role of organizational support practices. *Journal of World Business, 36*(4), 389–402.
88. Lazarova, M., & Cerdin, J. (2007). Revisiting repatriation concerns: Organizational support versus career and contextual influences. *Journal of International Business Studies, 38*(3), 404–429.
89. Selmer, J. (1999). Career issues and international adjustment of business expatriates. *Career Development International, 4*(2), 77–87.
90. Stahl, G. K., Miller, E., & Tung, R. (2002). Toward the boundaryless career: A closer look at the expatriate career concept and the perceived implications of an international assignment. *Journal of World Business, 37*(3), 216–227.
91. Dickmann, M., Doherty, N., Mills, T., & Brewster, C. (2008). Why do they go? Individual and corporate perspectives on the factors influencing the decision to accept an international assignment. *The International Journal of Human Resource Management, 19*(4), 731–751.
92. Hippler (2009). Why do they go? Empirical evidence of employees' motives for seeking or accepting relocation. *The International Journal of Human Resource Management*, *20*(6), 1381–1401.
93. Cole, N., & McNulty, Y. (2011). Why do female expatriates "fit-in" better than males? An analysis of self-transcendence and socio-cultural adjustment. *Cross Cultural Management: An International Journal, 18*(2), 144–164.
94. Scullion, H., & Brewster, C. (2001). The management of expatriates: Messages from Europe? *Journal of World Business, 36*(4), 346–365.
95. Suutari, V., Brewster, C., Riusala, K., & Syrjakari, S. (2013). Managing non-standard international experience: Evidence from a Finnish company. *Journal of Global Mobility, 1*(2), 118–138.
96. Câmara, P. B. (2008). *Gestão de pessoas em contexto internacional*. Braga: Dom Quixote.
97. Iellatchitch, A., Mayrhofer, W., & Meyer, M. (2003). Career fields: A small step towards a grand career theory? *International Journal of Human Resource Management, 14*(5), 728–750.

Index

C. Machado and J.P. Davim (eds.), *Managing Organizational Diversity*,
DOI 10.1007/978-3-319-54925-5

MIX
Papier aus verantwortungsvollen Quellen
Paper from responsible sources
FSC® C105338

If you have any concerns about our products,
you can contact us on
ProductSafety@springernature.com

In case Publisher is established outside the EU,
the EU authorized representative is:
Springer Nature Customer Service Center GmbH
Europaplatz 3, 69115 Heidelberg, Germany

Printed by Libri Plureos GmbH
in Hamburg, Germany